Personal Privacy in the Age of the Internet

Yair Oppenheim
Personal Privacy in the Age of the Internet

All rights reserved
Copyright © 2024 by Yair Oppenheim

No part of this publication may be reproduced, distributed, or transmitted in any form or by any means, including photocopying, recording, or other electronic or mechanical methods, without the prior written permission of the publisher, except in the case of brief quotations embodied in critical reviews and certain other noncommercial uses permitted by copyright law.

Published by Spines
ISBN: 979-8-89569-185-4

PERSONAL PRIVACY IN THE AGE OF THE INTERNET

THE INFLUENCE OF INFORMATION AND COMMUNICATION TECHNOLOGIES ON PERSONAL PRIVACY

YAIR OPPENHEIM

TRANSLATED BY
MARIA GERSHANIK

CONTENTS

Introduction

For two decades now, the issue of privacy has been the focus of public attention. Every other month, we seem to hear more outrageous news of hackers, companies or well-known organizations violating people's cyber privacy or stealing virtual identities to sell them. To name a few: in 2013, Edward Snowden leaked out highly classified NSA documents which proved that the United States security agencies had been monitoring thousands of individuals and organizations worldwide, invading and violating their privacy;[1] Cambridge Analytica Ltd. collected personal data from 50 million Facebook users without their consent and shared it to be improperly used in Donald Trump's 2016 presidential campaign;[2] hundreds of cases of virtual identity theft – mainly theft of credit card numbers from insurance companies, banks, and so on. These and other events have led numerous experts to declare the death of privacy[3] – so many times that it seems to never stop dying,

---

[1] Edward Snowden was a system administrator at the NSA, who in 2013 revealed thousands of classified documents proving how the US government had been systematically violating privacy all around the globe.

[2] As a result, Facebook founder and owner Mark Zuckerberg was called to testify before the US Congress to provide explanations and present a plan of action for improving Facebook users' privacy protection.

[3] Simon Garfinkel, *Database Nation: The Death of Privacy in the 21st Century*, 1st ed. (Sebastopol, California: O'Reilly, 2000), 1-10; Charles Sykes, "The End of Privacy", *Stanford* Vol. 61(1) (October 2008), 101-161; Reginald Whitaker, *The End of Privacy: How Total Surveillance is Becoming a Reality* (New York: New Press, 1999), 200; Jeffrey Rosen, *The Unwanted Gaze: The Destruction of Privacy in America* (New York: Vintage Books, 2000), 200.

or as Deborah Nelson puts it: "Privacy, it seems, is not simply dead. It is dying over and over again".[4]

Not everyone agrees that privacy is such a great matter of concern; some scholars believe the current preoccupation with privacy to be largely an expression of older Westerners' paranoia, bringing as evidence the ease with which people share personal information on social media. According to Calvin Gottlieb, "most people, when other interests are at stake, do not care enough about privacy to value it."[5] In contrast, Edward Snowden has famously said that "arguing that you don't care about the right to privacy because you have nothing to hide is no different than saying you don't care about free speech because you have nothing to say."[6] The heated debate around privacy, particularly personal privacy, is far from over as more and more individuals and authorities begin to realize the urgent need for regulation and supervision in this area. Some major examples include Mark Zuckerberg's testimony before the US Congress, as well as the European Union's new General Data Protection Regulation (GDPR)[7] designed to protect the personal information of individuals inside Europe and to prevent improper use of personal data. Either way, the issue of privacy, particularly various

---

[4] Deborah Nelson, *Pursuing Privacy in Cold War America* (New York: Columbia University Press, 2001), 4.

[5] Calvin C. Gottlieb, "Privacy: A Concept Whose Time Has Come and Gone", in: *Computer, Surveillance, and Privacy*, eds. David Lyon and Elia Zuriek (Minneapolis: University of Minnesota Press, 1996), 156.

[6] https://en.wikipedia.org/wiki/Nothing_to_hide_argument#:~:text=The %20nothing%20to%20hide%20argument,uncover%20their%20own%2 0illicit%20activities (Accessed 5/3/2024)

[7] Stephen Robert Massey, *The Ultimate GDPR Practitioner Guide: Demystifying Privacy & Data Protection* (London: Fox Red Risk, 2017), 1.

aspects of personal privacy,[8] persistently remains on the public agenda. However, a number of questions remain unclear: What is privacy as a concept? Where and to what does it apply? Where does the private sphere end and the public sphere begin? Where does privacy stand in relation to other concepts, such as freedom, autonomy and liberty? How to balance individual rights against the public's right to know? Must a state reveal potential threats to its security? And how should it all be regulated in this age of information and communication technologies that follow us wherever we go and whatever we do?

The aim of this book is to examine the influence of information and communication technologies (ICTs), big data, the Internet of Things (IoT) and artificial intelligence (AI) on personal privacy. ICTs store, analyze, fuse, and disseminate information to every individual who is active online, using abilities that did not exist until twenty years ago. An individual human is, at the same time, a subject that exists for itself and an element in a membership group (or groups). The individual as a concept only has meaning in the context of being part of a group (a society),[9] which is what allows us to define our individuality in reference to our membership groups. Thus, there is an apparent line that separates the individual from their membership groups, but it is flexible, frequently shifting, and time- and place-dependent. Within that line resides our personal privacy, and beyond it lies the public sphere. The individual can be regarded as an agent that interacts with the world[10] and carries its

---

[8] Personal privacy is the type of privacy that applies to individual humans, unlike general privacy, which also applies to organizations, corporations and agencies.

[9] Georg Simmel, "How is Society Possible?", *American Journal of Sociology* Vol. 16(3) (November 1910), 372-391.

[10] "Agency", *Stanford Encyclopedia of Philosophy*, first published Mon Aug 10, 2015.

personal privacy. Based on this duality, this book explores the various conceptions of privacy. Privacy is a continuous scale in which one end represents full personal privacy, and the other end represents complete lack thereof, rather than a dichotomy of having personal privacy versus not having any. Just like every other aspect of the relationship between individuals and society, the concept of personal privacy is culture-dependent and dictated by each particular culture (civilization) according to its values and social norms.

The type of society we are living in is what Alvin Toffler called the Third Wave[11] – a society in which knowledge and information are the primary resources and determinants of power. A Third-wave society makes extensive use of ICTs, with many human activities performed by virtual means – through personal computers or (mostly) smartphones – by identifying details and private personal information. That information is being stored, fused, analyzed and shared to support our way of life and serve us as individuals and as a society on the one hand, but on the other hand, to serve the economic and political interests of governments, conglomerates and social networks. The Third Wave civilization is new[12] and rapidly changing[13] – at a pace that humanity finds hard to follow – and is responsible for much of the unclarity regarding the concept of personal privacy. In my opinion, the cause of this unclarity is the fact that both the concept itself and the social norms related to it were formed for

---

[11] Alvin Toffler, *The Third Wave* (New York: Morrow, 1980), 17.

[12] Toffler, *The Third Wave*, 154: "From the mid-1950's it became increasingly apparent that these industries [the Second Wave industries – Y.O.] were backward and waning in the industrial nations."

[13] See Moore's law about computer chip performance doubling roughly every 2 years.

and based on a Second Wave culture and, as such, are unfit for a Third Wave society.

Many attempts have been made to redefine personal privacy, but none of them have been adequate.[14] In this book, I will use all those various definitions to extract the basic components of privacy and how they can be translated into categories of information.[15] In other words, I will **replace the discussion of personal privacy with a discussion of personal privacy information**. From there, I will propose a new distinction between deep personal privacy information and general personal privacy information. To help me map their place in the universe, I will use Karl Popper's theory of "Three Worlds"[16]: World 1 – the

---

[14] A partial list of scholars who have attempted to provide definitions for privacy:

Ferdinand D. Schoeman, "Privacy: philosophical dimensions", in: *Philosophical Dimensions of Privacy: An Anthology*, ed. Ferdinand D. Schoeman (Cambridge: Cambridge University Press, 1984), 3.

Graeme Laurie, *Genetic Privacy: A Challenge to Medico-Legal Norm* (Cambridge: Cambridge University Press, 2002), 6.

Soraj Hongladarom, *A Buddhist Theory of Privacy* (Singapore: Springer Singapore, 2016), 16.

Ruth Gavison, "Privacy and the Limits of Law", *The Yale Law Journal* Vol. 89, No. 3 (January 1980), 421.

Daniel J. Solove, *Understanding Privacy* (London: Harvard University press, 2009), 1-2.

Helen Nissenbaum, *Privacy in Context: Technology, Policy, and the Integrity of Social Life* (Stanford, Calif.: Stanford Law Books, 2010), 16.

[15] Despite Solove's view that it cannot be done: "The theory's focus on information, [...] it excludes those aspects of privacy that are not informational." – Solove, *Understanding Privacy*, 25.

[16] Karl Popper, "Three Worlds – The Tanner Lecture on Human Values", delivered at the University of Michigan (April 7, 1978). Available at: https://tannerlectures.utah.edu/_resources/documents/a-to-z/p/popper80.pdf (Accessed 3/3/2024)

physical world of natural objects and processes; World 2 – the mental or psychological world; and World 3 – the realm of the products of human thought. Their main characteristics are summarized in the following table:

	Contents	Status	Examples
World 1	Physical objects	Objective	Table, chair, steel
World 2	Mental experiences	Subjective	Pain, joy
World 3	Knowledge	Objective	Mathematics, Biology

Based on these definitions, I will place deep personal privacy in World 2 and general personal privacy in World 1 and 3.

The aforementioned distinctions allow us to draw a clear line around deep personal privacy and argue that until the age of ICTs, it has never been place-, time- or culture-dependent and is, therefore, a basic human need – unlike general personal privacy, which is culture-dependent and may be negotiated. Moreover, they show us that while ICTs are affecting both types of privacy, threatening to expose them and make them public, **the greatest change in this area is that ICTs allow – for the first time in history – the exposure of deep personal privacy**.

All the aforementioned issues will be discussed in this book in the following order:

Chapter 1: An overview of the information and communication technologies relevant to the issues discussed in this book, the ways they are used in processes that are part of our everyday lives, and how they specifically affect our personal privacy.

Chapter 2: The origins and roots of the need for personal privacy. This chapter explores the concept of privacy and the extent to which people need it and tries to answer the question of whether it can be called a fundamental human right.

Chapter 3: Personal privacy as a cultural phenomenon. In this chapter, I use a number of historical case studies to show how the conception of personal privacy is constructed by each culture and to support the argument made in Chapter 2 that human society cannot exist without personal privacy.

Chapter 4: Personal privacy in the modern era. This chapter describes the philosophical, cultural, social and economic base of the Western liberal notion of personal privacy.

Chapter 5: The Industrial Revolution, individualism and the new definition of personal privacy. This chapter provides an overview of the main characteristics of the modern industrial (pre-information age) society and shows how the social norms and regulations related to privacy derive from those characteristics.

Chapter 6: Personal privacy – identity and private sphere. This chapter discusses the definition of personal privacy and its various components and introduces the terms "deep personal privacy" and "general personal privacy."

Chapter 7: The paradigm of personal privacy. This chapter frames the various conceptions of personal privacy shaped by the democratic Western civilization[17] under a common **paradigm**.

Chapter 8: Personal privacy information. In this chapter, I explain how privacy information is associated with each of the basic

---

[17] The following chapters include a discussion whether the information revolution has created a need for a new paradigm of personal privacy.

components of privacy, how it meets the criteria that define an entity as information, how it fits into the paradigm of personal privacy described in Chapter 7, and how the components of personal privacy can be broken down into distinct categories of information. This brings us to the main argument of this chapter: that any discussion of personal privacy can be replaced with a discussion of personal privacy information – a concept that is clear, measurable and quantifiable.[18]

Chapter 9: Personal privacy in the age of ICTs. A description of the information revolution (the Third Wave)[19] and its role in the formation of the network society, the network market and the network human, followed by a discussion of the resulting change in the conception of personal privacy in liberal Western society, the impact of that conceptual change on what it means to be a human individual in a network society, and the impact of one's role as a node in the new social network on one's personal privacy and social interactions.

Chapter 10: Personal privacy information is bound to be dispersed. This chapter introduces a few "laws of nature" due to which information in general, and personal information in particular, is bound to be dispersed, and shows how personal privacy is a complex system subject to the law of entropy and how ICTs are accelerating the entropy increase, bringing personal privacy to the brink of chaos.

Chapter 11: Toward a new paradigm of personal privacy. In this chapter, I explain how the information revolution has cracked the paradigm of personal privacy described in Chapter 7 and why those

---

[18] Claude E. Shannon, "A Mathematical Theory of Communication", reprinted with corrections from *The Bell System Technical Journal* Vol. 27 (October 1948), 379-423.
[19] Toffler, *The Third Wave*, 17.

cracks cannot be mended within the current paradigm. Finally, I draw an outline of a new paradigm of personal privacy.

Chapter 12: A summary and a discussion of options and potential mechanisms of action for protecting personal privacy, particularly deep personal privacy, in the age of ICTs.

This book addresses every aspect of personal privacy, focusing on how it is influenced by information and communication technologies. The influence of ICTs is forcing us to find new answers to the questions of our self-conception (who we are), our mutual interactions (how we socialize), our conception of reality (our metaphysics), and our interactions with and impact on reality (our agency). **Each of those questions affects our conception of personal privacy and how we live it out in the world.**

Chapter 1

The Age of ICTs

The Components of ICTs

The hardware that is the basis of information and communication technologies (ICTs) is electronics and computers, which have been advancing at an increasing rate since World War II.[20] At the core of the hardware is an electronic processor designed according to the von Neumann (Princeton) model[21] composed of a central processing unit (CPU), memory, a control unit, and input and output mechanisms. The memory stores data and algorithms (work procedures). The role of the CPU is to run specific algorithms from the memory, according to which it processes the data stored in the memory. The input and output mechanisms are used for communication with the outside world; through them, the processor receives instructions (how and what to process) and reports the processing results. The control unit synchronizes the work of the computer's basic components.

---

[20] Isaac Ben-Israel, Lior Tabansky, "Mabat beintkhumi al etgarey ha-bitakhon be-idan ha-meyda" [A multidisciplinary overview of security challenges in the information age] (in Hebrew), *Tzava ve-estrategia* Vol. 3 (December 2011), 19-32.

[21] Sara Folk, Dan Aharoni, *Mavo le maarakhot makhshev ve-assembly* [Introduction to Computer Systems and Assembly] (in Hebrew) (Tel Aviv: Centre for Educational Technology, 2006), 13.

Hardware capacity is growing rapidly in accordance with Moore's law[22] – an observation made in 1965 by Gordon Moore that the number of transistors in an integrated circuit (IC, or microchip) doubles about every two years. Moore's prediction has come true and still applies today, resulting in exponential growth in computer and processor capacity, which allows us to use electronic processing technologies for an ever-growing variety of applications – including control of missiles launched into space, use in weapons, creation of smart communication devices such as smartphones, and control of virtually any device used by humanity. The cost of a single processor has significantly reduced since Moore's times, making processors affordable to practically everybody. These two processes are mutually accelerating, contributing to the wide distribution of processors.

This development has created a new type of space: the cyberspace. Cyberspace is a human creation that has joined the physical spaces in which humankind has been active since the dawn of civilization – land, sea, air and outer space – enhancing the interactions between them and human capacities within each.

Other important consequences of the rapid development of computer hardware were the ability to create giant databases at a low cost, resulting in the formation of big data, and the emergence of the Internet communication system, which covers the entire world – sometimes by means of fiber optic and copper communication cables, and sometimes by wireless transmission of electronic signals (Wi-Fi connection). During the last two decades, it also gave rise to what is called the Internet of Things (IoT)[23]: the ability to communicate with

---

[22] Ethan R. Mollick, "Establishing Moore's law", *IEEE Annals of the History of Computing*, Vol. 28(3) (July 2006), 62-75.
[23] Eran Alroy, "Ma ze, le-azazel, 'ha-Internet shel ha-dvarim'?" [What

various devices, such as refrigerators and cars, using Internet processors, sensors and cameras. The IoT provides a new way for people to communicate with each other and with various devices and applications. It allows devices to communicate with each other and exchange information and instructions with increasing autonomy. In 2014, Gartner made a prediction (which came true) that by the end of 2020, the IoT would grow to 26 billion connected devices worldwide.[24] According to McKinsey, the global IoT market will reach $620 billion by 2025.

Thus, one of the major consequences of the rapid development of the hardware described above was the global Internet communication network, whose infrastructure includes the World Wide Web, telecommunications networks, computer systems, processors, microchips and controllers, and which traffics enormous amounts of information and data. In addition, it created a base for big data and the advent of artificial intelligence (AI), particularly deep learning applications.

The Software of ICTs

ICT software is comprised of several layers. The first layer is the operating system (OS) – the software program that manages the processor and the various devices. Every OS has several versions. The most popular operating systems for computers are Microsoft Windows,

---

the heck is 'the Internet of things'?] (in Hebrew), *Anashim u Makhshevim*, October 2014. Available at: https://www.pc.co.il/editorial/168117/ (Accessed 21/9/2020)

[24] https://www.finyear.com/Gartner-Says-the-Internet-of-Things-Installed-Base-Will-Grow-to-26-Billion-Units-By-2020_a27901.html (Accessed 5/3/2024)

Linux, Unix-based systems and IBM System/360. For smartphones, Apple uses the iOS operating system, while most of its competitors use Android. The OS of PLCs (programmable logic controllers) are communication protocols that define how the network components communicate with each other. The most common communication protocol is TCP/IP – the Internet communication protocol suite.

The second layer is the programming language – a set of syntax and semantic rules that define a formal language. The programming language in which a program is written dictates the performance of the algorithmic process and the setting tools at the programmer's disposal. Thus, programming languages are a way for humans to communicate with computers and translate human instructions to the computer's machine language. Many programming languages have been developed over the years, with different emphases and adjustments made to comply with different OS. Programming languages can be roughly divided into two groups: assembly (ASM) languages, which are very close to machine languages, and high-level programming languages, which use more abstract concepts and can achieve the required performance of the algorithm even with little understanding of the hardware it is running on.

App is the name given to a software application usually activated by clicking or tapping on an icon in a browser or on the smartphone screen. Still, in fact, an app consists of one or more software programs designed to perform a particular task or set of tasks with a specific purpose. Global software infrastructures such as Facebook, Google Chrome, and Oracle are also apps. This book focuses on apps such as cookies,[25] which are planted in personal computers and

---

[25] For an explanation on how cookies work, see:
Emmanuel Benoist, "Collecting Data for the Profiling of Web Users", in: *Profiling the European Citizen: Cross-Disciplinary Perspectives*, eds.

smartphones to collect information and monitor and learn the user's online movement and behavior.

The term "information and communication technologies" (ICTs) includes computing technologies[26] in which hardware, software and data are combined to create an entity with unique characteristics. That includes the World Wide Web[27] – a global system of interlinked webpages stored on servers, each of which has a unique address called URL (uniform resource locator). The information is stored on those pages, which are built using a programming language called HTL (hypertext markup language) and transferred using a communication protocol called HTTP (hypertext transfer protocol), which belongs to the application layer of the OSI or TCP/IP model. HTTP servers are the main content servers on the Internet network, and browsers are the most common client software.

Big data[28]: A computer can store an enormous amount of data, both structured and unstructured. Those huge databases are continuously fed new commercial and other data and are arranged in a way that allows very large amounts of data to be quickly processed, analyzed, cross-referenced and turned into information that can be used by the database controller to make better commercial (or other) decisions.

Mireille Hildebrandt and Serge Gutwirth (Dordrecht: Springer, 2008), 169-184.

[26] Including mathematical approaches such as data mining and neural networks.

[27] David Gourley and Brian Totty, *HTTP: The Definitive Guide* (Sebastopol, California: O'Reilly, 2002), 596.

[28] Douglas Laney, "Application Delivery Strategies", *META Group Inc.*, Feb 2001.

Deep learning technology[29]: A method in AI and machine learning that uses multi-layered neural networks and statistical methods to assess the presence and strength of specific properties. This book focuses on how deep learning technology can be used to violate personal privacy.

Cloud computing[30]: A model that allows easy, on-demand online access to a shared pool of computer system resources – mainly computing power, data storage and connectivity. This book shows how cloud computing can be used to cross-reference data from different sources in order to make conclusions, violating personal privacy.

To wrap up the last two sections, we can see that ICTs have become world-encircling interdependent technologies that lean on a combination of hardware and software.

The Cultural, Social and Economic Implications of ICTs

Many scholars believe the cultural, social and economic implications of ICTs to be no less than a revolution. One of them is Alvin Toffler, who says in his book *The Third Wave* that the information revolution has extended the sphere of technology and given rise to a new civilization that is "so profoundly revolutionary [...] that it challenges all our old assumptions".[31] Toffler looks at the development of human civilization as a succession of rolling waves of change that are not separate but following each other, each one rising before the previous has subsided.

---

[29] https://www.ibm.com/topics/deep-learning (Accessed 12/3/2024)

[30] David Chappell, "A short introduction to cloud platforms: An enterprise-oriented view", 3-4.

[31] Toffler, *The Third Wave*, 18; Ben-Israel and Tabansky, "Mabat beintkhumi al etgarey ha-bitakhon be-idan ha-meyda", 21.

He divides history into **The First Wave** – the transition from hunter-gatherer society to agricultural society, which began roughly 8000 years ago and gradually spread across the entire globe; **The Second Wave** – the industrial phase, which started around 300 years ago and gave us factories, trains, airplanes and polluting energy; and a rising **Third Wave** – a technological revolution which began around the mid-1960s and is currently creating a new form of society based on knowledge, sophistication and information-based economy with decentralized production.[32] Toffler's "waves" theory can be summarized as follows:

	First Wave	Second Wave	Third Wave
Main characteristic	Agriculture	Industry	Information[33]
Primary resource(s)	Land	Machines	Data and knowledge

Each wave's main characteristics and primary resources dictate who will have the economic and social power – namely, the people who own the primary resource of that wave. The tools that best symbolize the First Wave are the plough and the scythe, and the people in power were the landowners and those who controlled water and access to it. The symbol of the Second Wave is the factory assembly lines, and the people who controlled them – the industrializers and factory owners – had the power. The symbol of the Third Wave – the age of ICTs – is the computer,

---

[32] Toffler, *The Third Wave*, 25-27.
[33] For a definition of information, see Chapter 8.

and the people in power are the ones who own the ICTs and control the information and knowledge.[34]

According to Manuel Castells,[35] a revolution is characterized by the type of fuel that powers it. The Industrial Revolution was powered by fossil energy, while the Information Revolution is powered by information, which – though it is not literally a fuel – is an essential resource for ICTs. Thus, Castells also defines it as a revolution (or a "wave," to put it in Toffler's terms). According to him, the influence of ICTs is so great that anything can be translated into information. Thus, information connects everything and is connected to everything. Castells also points out the information revolution's incredibly fast pace.

Luciano Floridi also perceives the rise of ICTs as a revolution and calls it "a fourth revolution in our self-understanding, brought about by computer science and ICT applications, after the Copernican, the Darwinian and the Freudian ones."[36] According to Floridi, each of those three revolutions changed the way we perceive ourselves. The Copernican revolution has shown humankind that our planet is not the center of the universe and not everything revolves around human beings. The Darwinian revolution showed us that we are not being created in God's image. Still, the result of a long and complex process of evolution that began with very simple unicellular organisms and which clearly does

---

[34] E.g., Bill Gates – the owner of Microsoft and inventor of Windows PC operating systems; Facebook owner Mark Zuckerberg; Google founders Larry Page and Sergey Brin; Jeff Bezos, the owner of Amazon e-commerce and cloud services company; and other prominent personalities in the ICT industry.

[35] Manuel Castells, *The Rise of the Network Society* (Oxford: Wiley-Blackwell, 1996), 31.

[36] Luciano Floridi, "The Informational Nature of Personal Identity", *Minds and Machines*, Vol. 21 (4) (November 2011), 549-566.

not end with us, and so may in the future produce beings superior to humans. The Freudian revolution separated the human self into at least three basic elements, showing us that the human soul is not unitary. The ICT revolution regards the human self as something that can be translated into pieces of information. Each of the theories that consider the age of ICTs to be a revolution that gives birth to a new reality includes great implications for personal privacy, which will be discussed in the next chapters.

Even if we take a neutral stand regarding the powerful changes that are taking place in the age of ICTs, I think we can all agree that ICTs have completely transformed many of our Western lifestyle behaviors, including: e-commerce – as a large and still growing section of commerce now takes place online via global companies such as Amazon and Alibaba, which specialize in online commerce; consuming information and entertainment digitally through news companies like Fox News, online streaming services like Netflix, and Kindle e-books; personalized DNA-based medicine; increasing amounts of information-based industries and services that can be received at home, in our private space; use of online search engines like Google and online databases like Wikipedia to find information, replacing traditional encyclopedias as the primary source of knowledge and challenging the status of educational institutions; making and maintaining friendships and social relationships via social media applications like WhatsApp, Facebook and TikTok; and using programs like Waze to help us navigate in the physical world.

All these uses we make of ICTs have wide-ranging implications for our personal privacy. You simply cannot make them without providing extensive information about yourself. Apps need to get our personal information in advance before we even begin using them, so they collect it both before and during use. The ICT components reviewed above are used to collect information about physical persons and to continuously monitor users' movement by collecting and storing the footprints they leave in the physical and virtual space in giant databases (big data pools). Vast amounts of information are collected through our smartphone – when we are active (using apps), but also when we are not actively using it – and through a wide range of devices and sensors such as CCTV cameras (commonly used in "smart cities"[37]), web cameras, and geolocation satellites that can locate a particular person in the physical space at any given moment. "Cookies"[38] planted in our PCs and smartphones collect information and monitor and identify our movements and behavior in the virtual space.

Information collection and surveillance are practiced all over the world by government organizations[39] as well as by commercial companies using commercial applications. The collected data undergo "profiling"[40] – the creation of a data pool about an individual, an organization or a physical or Internet entity.[41] In this process, data from different sources are analyzed using various algorithms, including deep learning, to make a profile of that individual or entity. The profile is then

---

[37]   Sam   Musa,   "Smart   City   Roadmap", https://www.academia.edu/21181336/ (Accessed Jan 2016)

[38] Benoist, "Collecting Data for the Profiling...", 169-184.

[39] As exposed by Edward Snowden in June 2013. See:
Michael Calderone, "Washington Post Began PRISM Story Three Weeks Ago, Heard Guardian's 'Footsteps', "*Huffpost*, 7 June 2013. https://www.huffpost.com/entry/washington-post-prism-guardian_n_34 02883 (Accessed 6/7/2013)

[40] Mireille Hildebrandt, "Defining Profiling: A New Type of Knowledge?", in: *Profiling the European Citizen*, eds. Mireille Hildebrandt and Serge Gutwirth (Dordrecht: Springer, 2008), 17-45.

[41] An Internet entity is a virtual entity, such as a Facebook user, while a physical entity is a real-life person or machine.

used to predict various behavior patterns, such as commercial preferences, social habits, and even political opinions, with a high level of accuracy. Personal information that relates to every aspect of our lives is being collected, stored, analyzed and shared as an inherent part of this surveillance process. After the information is processed, it becomes a commodity to be traded between various consumers of information, government and private alike. In the age of ICTs, the individual is under constant and extensive surveillance, and this has a lot of implications for the question of personal privacy.

Another characteristic of cyberspace is the "small world problem": human society has become a network with six degrees of separation, which means that any two people are six or fewer social connections away from each other.[42] Marshall McLuhan, in his book *Understanding Media*, describes it as follows: "After three thousand years of specialist explosion and of increasing specialism [...], our world has become compressional by dramatic reversal. As electrically contracted, the globe is no more than a village. Electric speed in bringing all social and political functions together in a sudden implosion has heightened human awareness of responsibility to an intense degree".[43] The age of ICTs has created a world that is small and compressed, like a village where everybody knows each other, in which social and personal distance is minimized, and the anonymity element of personal privacy seems to have disappeared.

In the new cyberspace, time and place become insignificant as information moves through the Internet network at the speed of light. In

---

[42] Jeffrey Travers and Stanley Milgram, "An Experimental Study of the Small World Problem", *Sociometry* Vol.32 (4) (Dec. 1969), 425-443.

[43] Marshall McLuhan, *Understanding Media: The extensions of man* (London and New York: McGraw-Hill, 1964), 7.

the human experience, everything is happening here, and now, all the residents of the global village are present and taking part in the discussions and commercial activities in the "town square," regardless of their physical position in space and time. Social media apps use these physical properties, reducing the degrees of separation in our social network and increasing the amount of information that flows through it.

Chapter 2

Our Need for Personal Privacy

According to Glenn Negley, the question of privacy has often been overlooked by philosophers,[44] but lately, it has been given much attention. There is no consensus among philosophers, politicians and lawmakers on whether privacy is a natural human right. Few philosophers would agree that the right to privacy is "natural," i.e., universal, fundamental and not dependent on time or place. In modern societies, the question of privacy is receiving more attention than it has in the past, but according to Negley, that does not make it a natural right.

For the last couple of decades, personal privacy has been the focus of public discussion. Every other month, we seem to hear more outrageous news of well-known organizations or companies violating people's cyber privacy or of hackers stealing others' virtual identities to sell them. To name a few: in 2013, Edward Snowden[45] leaked out highly classified NSA documents which proved that the United States security agencies had been monitoring thousands of individuals and organizations worldwide, invading and violating their privacy; Cambridge Analytica Ltd. collected personal data from 50 million

---

[44] Glenn Negley, "Philosophical views on the value of privacy", *Law and Contemporary Problems* 31 (Spring 1966), 319-325.

[45] Edward Snowden was a system administrator at the NSA, who in 2013 revealed thousands of classified documents proving how the US government had been systematically violating privacy all around the globe.

Facebook users without their consent and shared it to be improperly used in Donald Trump's 2016 presidential campaign; hundreds of cases of virtual identity theft – mainly theft of credit card numbers from insurance companies, banks, and so on. Privacy is mostly regarded as something culture-dependent or even as a basic need for maintaining the social contract between individuals and between individuals and society. Corporate privacy, e.g., the privacy of a bank, which includes its confidential information and corporate secrets, can be regarded as an extension of personal privacy – which applies to individuals – to organizations, which are groups of people.

The Individual as an Agent of Privacy

Before we begin to talk about the roots of personal privacy, we must first define two basic concepts: "individual" and "society." Without them, there can be no discussion of privacy, which is their derivative. The Oxford English Dictionary defines "individual" as "a person considered separately rather than as part of a group," "a single member of a class of things," or "a person who is original and very different from others." So, linguistically speaking, an individual is always considered in reference to a group. As such, every individual has properties that are unique to it, as well as properties it shares with the rest of the group(s) to which it belongs. Such a definition of an individual in reference to a group indicates that there can be no "individual" without a membership group. In the case of human individuals, sociological research argues that without a society – a membership group – the "individual" has no meaning.[46] Every individual has a different and unique inner core that

---

[46] Simmel, "How is Society Possible?", 372-391.

cannot be entirely known to others, be it other individuals or society as a whole. Because the individual only has meaning in the context of being part of a group (a society), the individual cannot, by definition, completely blend and merge with the other. Even the religious person who seeks to be completely encompassed by the divine being must retain some sort of self-existence, of being other than God, or their strife to be one with God will lose its meaning. Georg Simmel goes on to say that "only through their generic and societary merging were the factors produced in the synthesis of which in turn the ostensible individuality may consist. On the other hand, we know ourselves as a member of society". Being part of a group is what allows one to define one's individuality in reference to one's membership group or groups. Thus, privacy is the line that separates the individual from their membership groups. That line is flexible, frequently shifting, and has two different sides: from within, the individual defines themself in reference to their membership groups (society);[47] from without, the membership groups define the individual as their member.

The individual human can be defined as an agent that interacts with and affects the reality around it. Every agent must meet the following requirements:[48]

- Individuality – Distinguishable from its environment;
- Interactional asymmetry – Being a source of action more than being acted upon;
- Normativity – Acting intentionally and according to norms.

---

[47] Family is also a reference group for human individuals.
[48] Xabier E. Barandiaran, Ezequiel Di Paolo, Marieke Rohde, "Defining Agency: Individuality, Normativity, Asymmetry, and Spatio-temporality in Action", in: *Adaptive Behavior* Vol. 17 (5) (October 2009), 367-386.

These three requirements – individuality, normativity and interactional asymmetry – are necessary but not sufficient conditions for agency; all three must be present at the same time and be interconnected. An individual human fulfills all three requirements at the same time: has an individual distinguishable identity, is a source of action, and acts according to norms.[49]

Human agents act in a social network of mutual influences.[50] In terms of privacy, a human agent fulfills the three requirements: (1) Carries its own privacy, distinguishable from that of others; (2) Is a source of action, defining the contents and boundaries of its privacy while interacting with the social network around it which defines and delimits its privacy from without; (3) Defines its privacy within the limits of social norms and regulations.

Based on this duality of the individual being at the same time a subject that exists for itself and a member of social groups, we can say that the concepts of "individual," "private," and "personal privacy" have no meaning except in the context of the group or groups to which the individual belongs, because privacy means distinguishing one member from its membership group (society). Therefore, privacy cannot exist without publicness, and vice versa. There is a dialectic synthesis between the definitions of the individual as "a person considered separately" and as "a single member of a class." The two are inseparable, and together, they create the unitary individual. Simmel sums it up very clearly:

"The essential thing, however, and the meaning of the particular sociological apriori which has its basis herein, is this, that between

---

[49] "Agency", *Stanford Encyclopedia of Philosophy*, first published Mon Aug 10, 2015.

[50] Albert Bandura, "Social Cognitive Theory: An Agentic Perspective", *Asian Journal of Social Psychology*, Vol. 2 (April 1991), 21-41.

individual and society, the Within and Without are not two determinations which exist alongside each other − although they may occasionally develop in that way, and even to the degree of reciprocal enmity − that they signify the whole unitary position of the socially living human being. His existence is not merely, in the subdivision of the contents, partially social and partially individual, but it stands under the fundamental, formative, irreducible category of a unity, which we cannot otherwise express than through the synthesis or the contemporariness of the two logically antithetical determinations − articulation and self−sufficiency, the condition of being produced by, and contained in society, and on the other hand, of being derived out of and moving around its own center."[51]

Despite this description of the individual as a unitary and seemingly undividable basic unit, I will later offer an approach that breaks personal privacy down into its components and regards the individual as an entity comprised of smaller elements (e.g., body, mind, etc.). Now, to complement our definition of an individual as a human agent that carries its personal privacy, we must also define the environment in which that agent acts, i.e., society. The Oxford English Dictionary defines it as "people in general, living together in communities" or as "a particular community of people who share the same customs, laws, etc.". In other words, society is the human environment in which the individual is an agent. It is the envelope that surrounds the individual, its boundaries and power determined through conflict and struggle with the individual, but it also complements the individual. As suggested by the definition, a society is not a unitary

---

[51] Simmel, "How is Society Possible?", 386-387.

entity; it is comprised of many different groups, some of which the individual is a member.

Every society has unique characteristics based on the individuals that comprise it and the relationships between them, which are a derivative of the social norms, laws, and customs. In Simmel's words: "the individuality of the individual finds a position in the structure of the generality, and still more that this structure in a certain degree, in spite of the incalculability of the individuality, depends antecedently upon it and its function."[52]

Between individuals and society, there is intrinsic tension and a constant struggle over the boundaries of each. Neither can exist without the other, and therefore, that struggle is never a zero-sum game. **The relationship between individuals and society is conflictive and complementary at the same time because one cannot exist without the other.** This tension also creates a balance between personal privacy and publicness. Personal privacy is the opposite of publicness, but it is also opposite to other individuals because it is threatened not only by society but by other individuals as well. It is the interest of human agents to reduce the privacy of other human agents. The balancing system between the individual and other individuals, and between the individual and society (as a group of other individuals), is dynamic and affected by social norms and regulations and by the technological environment. **Privacy can be defined as the inherent conflict between individual agents and other parties such as governments, organizations and other groups, but also as a complementary relationship.**[53]

---

[52] Ibid., 391.

[53] I will elaborate further on this conception of privacy as an inherent conflict, and its implications.

According to Bruce Schneier, the most common misconception regarding privacy is that it has to do with something bad that you want to hide.[54] I tend to agree that privacy does not necessarily mean having something to hide: we are not doing anything wrong when we shower, have sex or express our feelings and emotions.

I would like to make a conceptual distinction between the terms "private" and "personal." **"Personal"** means something that belongs to or is completely identified with a specific individual (even if it is not unique to that individual), such as body appearance, thoughts and actions. On the other hand, **"private"** and **"personal privacy"** refer to all the things that should be inaccessible to others and are univalently identified with a specific individual. This applies not only to abstract states but also to physical things and facts, such as the appearance of private body parts. When one comes to examine the relationship between **personal** and **personal privacy**, it seems that **personal privacy** is included in the definition of **personal**. Our thoughts, for example, belong in the sphere of personal privacy until we choose to express them and share them with others. After that, they remain our personal thoughts, though they are no longer private. The line between **personal** and public passes wherever the **personal** is taken out of the individual's control and transferred to an external entity, e.g., when people are told what to think (brainwashed) or when an individual's property is used by others without permission.

---

[54] Bruce Schneier, *Data and Goliath: The Hidden Battles to Collect Your Data and Control Your World* (New York: W. W. Norton and Company, 2016), 125.

The Origins of the Need for Personal Privacy

According to Alan Westin, even though privacy is a distinctly human need, its roots can be traced back to a primitive mechanism that is common to humans and other animals that require "privacy" from other members of their group.[55] The most explicit manifestation of the need for privacy in humans and other animals is marking a private territory that is not to be shared with other individuals of the same species. There is much evidence of animals fighting for territory[56] and trying to defend the territory they had marked as their own against intrusion (invasion) by members of their own species. Pipits jealously defend the six-foot radius around them from other pipits, except during the breeding period.[57] The three-spined stickleback guards an invisible water wall around itself and attacks any other stickleback that swims into it. A study in rats has shown that rats in a crowded cage would fight to the death over food and water, even when those are in abundance, and that lack of physical space that would allow separation between individuals is associated with a decrease in the number of births.

The need for a private territory, which is common to all animals, including humans, is manifested in maintaining distance from other individuals. Adam D. Moore calls privacy "a necessary condition for

---

[55] Alan Westin, "The origins of modern claims to privacy", in: *Philosophical Dimensions of Privacy: An Anthology*, 56-59; Seymour Lieberman, "The Effect of Change in Role on the Attitudes of Role Occupant", in: *Human Behavior and International Politics*, ed. J. D. Singer (Chicago: Rand-McNally, 1965), 155-157.

[56] H. Eliot Howard, *Territory in Bird Life* (New York: E. P. Dutton and Co., 1920), 1-4.

[57] Westin, "The origins of modern claims to privacy", 56.

human well-being."[58] Overpopulation, which would mean a reduction of personal space, may lead to a suicidal reduction, as in the case of lemmings marching into the sea and drowning.[59] This argument is supported by studies showing that privacy is a fundamental norm in every human culture.[60] It may have different manifestations in different cultures, but the basic need to have privacy is universal. Therefore, there is a clear answer to the question of whether privacy has the justification of being a fundamental human right or whether it is entirely culture-dependent.[61] According to Moore, it is a basic human need manifested through social norms.[62] According to Soraj Hongladarom, privacy stems from the human need for autonomy and subjective existence.[63] Autonomy and subjective existence are necessary for independent thought, and that is one of the most important justifications for privacy: "It seems clear that the ability to think for themselves is closely related to the individual's ability to have their private sphere of thoughts and feelings. It is presumably in virtue of their having their private, individual spheres that individuals are autonomous, thus becoming candidates for moral worth and dignity." In other words, Hongladarom believes privacy is a basic need, says it is necessary because it allows people to think for themselves, and emphasizes the human need for autonomy as a precondition for self-fulfillment.

---

[58] Adam D. Moore, "Privacy: Its Meaning and Value", *American Philosophical Quarterly*, Vol. 40 (3) (July 2003), 215-227.

[59] The lemming (Lemmus) is a small rodent from the subfamily Arvicolinae.

[60] Moore, "Privacy: Its Meaning and Value".

[61] Hongladarom, *A Buddhist Theory of Privacy*, 31-32.

[62] Moore, "Privacy: Its Meaning and Value".

[63] Hongladarom, *A Buddhist Theory of Privacy*, 14.

Ferdinand Schoeman believes human dignity – which includes the right to privacy – to be an inherent right that stems from a person's being a person rather than a tool for achieving external ends.[64] Every individual has a core that must not be exploited by others, even for good social purposes. Privacy protects us and prevents us from becoming a tool for achieving social goals by separating our personal goals from social ones. It allows people to live without supervision, with respect for their personal goals. Barrington Moore Jr. says the claim to privacy is rooted in the need for a private physical space and makes a clear distinction between virtual emotional space and territorial physical space.

As we can see, even though different philosophers put their emphasis on different aspects, there exists a general consensus that the claim to privacy stems from a fundamental human need that is manifested in every culture in one way or another. According to Milton R. Konvitz, "Once a civilization has made a distinction between the "outer" and the "inner" man, between the life of the soul and the life of the body, between the spiritual and the material, between the sacred and the profane, between the realm of God and the realm of Caesar, between church and state, between rights inherent and inalienable and rights that are in the power of government to give and take away, between public and private, between society and solitude, it becomes impossible to avoid the idea of privacy by whatever name it may be called – the idea of a "private space in which man may become and remain 'himself.' "[65] In other words, privacy is where our selfhood is born and maintained. The

---

[64] Ferdinand D. Schoeman, *Philosophical Dimensions of Privacy: An Anthology*, 4.

[65] Milton R. Konvitz, "Privacy and the Law: A Philosophical Prelude", *Law and Contemporary Problems* 31 (2) (Spring 1966), 272-280.

question of what exactly "selfhood" is has many different answers. I will mention two psychological approaches described by Ferdinand Schoeman:

According to the first approach, every individual human has an inner nucleus, a core self that we never reveal in full but choose to show some parts and to keep others hidden to a greater or lesser extent, depending on our needs and circumstances. Because all the behaviors of the same individual share a common core, that core can be inferred based on one behavior, allowing one to predict other behaviors.

The second approach, which is supported by lab studies, says there is no "atomic" inner nucleus; the self has many faces, which are manifested in different circumstances. One person can have multiple "selves" (identities) without having schizophrenia. Each behavior is independent of the other, and it will be easier to predict an individual's behavior based on the behavior of other members of the same culture in similar circumstances than on other behaviors of the same individual.[66]

Either way, both approaches assume some personal space around the individual. Westin calls the imaginary radius that defines a personal space "social distance."[67] Humans and other complex mammals use their five senses (sight, hearing, touch, smell and taste) to maintain it.[68] The idea of social distance is based on the philosophy of Georg Simmel, who describes the need to balance between self-revelation and self-concealment as a condition for proper social relationships between

---

[66] Schoeman, "Privacy and intimate information", in: *Philosophical Dimensions of Privacy: An Anthology*, 410-411.

[67] At the end of this chapter, I will propose a metric for calculating this distance, and replace the term "social distance" with "extent of non-knowledge of personal privacy information".

[68] Westin, "The origins of modern claims to privacy", 57.

individuals.[69] For Simmel,[70] the roots of social distance[71] lie in the information and knowledge we have about each other. The extent of familiarity or unfamiliarity between people is the sphere of personal privacy, which is the very thing that allows social relationships between individual humans.[72] A society in which people know too much about one another will not last long.[73] The limit on the amount of personal information that may be exposed to the public without ruining the social fabric can and should be negotiated. However, some social distance around each individual seems to be essential in order to maintain proper social relations.[74]

Michael Birnhack offers two approaches to personal privacy, both of which regard it as a protective envelope surrounding the individual: one is "privacy as access to the individual," which looks at that "envelope" from outside, and the other is "privacy as control."[75] According to both, penetration of the radius of social distance – such as

---

[69] Simmel, "How is Society Possible?", 372-391.

[70] Robert F. Murphy, "Social distance and the veil", in: *Philosophical Dimensions of Privacy: An Anthology*, 34-35.

[71] See below a proposed metric for calculating specific categories of social distance.

[72] Georg Simmel and Kurt H. Wolff, *The Sociology of Georg Simmel* (Glencoe, III: Free Press, 1950), 312.

[73] Murphy, "Social distance and the veil", 35.

[74] See the mathematical section for a way to calculate the "minimal radius" of social distance. As I will explain, this radius is culture-dependent.

[75] Michael D. Birnhack, *Merkhav Prati: Ha-zkhut la-pratiyut ha-ishit beyn mishpat le-technologia* [Private Space: The Right to Privacy, Law and Technology] (in Hebrew) (Ramat Gan: Bar-Ilan University, 2010), 92.

I will discuss these two approaches in further detail in the chapter about the definitions of privacy.

intruding into one's home without one's consent, making unpleasant smells or making loud noise – is considered a violation of privacy. The social distance mechanism is so deeply rooted that it triggers a reaction in animals – and, apparently, in humans too – even without objective reason (e.g., when one does not actually have to control a territory to defend vital resources) and will be manifested as aggression toward other members of the species whenever social distance is reduced due to overpopulation. Animals would kill each other or commit mass suicide, like the lemmings,[76] in order to reduce population density. Different cultures have developed different mechanisms for maintaining social distance. Robert F. Murphy describes the custom of Tuareg males to cover their faces with a veil, with only the areas around the eyes and nose remaining visible.[77] This custom is not dictated by the Saharan climate in which they live but marks the wearer's social status and helps maintain social distance between individuals. The veil allows one to conceal himself from others by hiding facial expressions that show his feelings and emotions and to control the extent of the revelation of those inner feelings. Murphy adopts Simmel's view of social distance as a necessary condition for any social relationship, which increases in direct ratio to the intimacy of the relationship. Even in the most intimate relationships, each individual must maintain some distance (privacy) in order to make the relationship possible. Social communication and social distancing are not mutually exclusive.

---

[76] Westin, "The origins of modern claims to privacy", 57.
[77] Murphy, "Social distance and the veil", 41-47.

Personal Privacy as a Conflict Between Individual and Social Needs

As we saw in the previous section, traditional discussions of personal privacy focus on the individual human's rights and needs in relation to society and on the individual's separation from society. As Richard Hixson puts it, "Generally speaking, the concept of privacy, as old as human history, tries to distinguish between the individual and the collective, between self and society. The concept is based upon respect for the individual, which has evolved into respect for individualism and individuality."[78]

Looking at personal privacy through the prism of individual rights and needs is natural, but it includes a perception of personal privacy as a threat to society. According to Garfield Benjamin, throughout human history, demands of privacy by marginal social groups (e.g., insurgents, persecuted religious groups such as Christians in Rome and Jews under the Spanish Inquisition, or homosexuals until the 1980s) have, in fact, been demands for freedom of action.[79] When their demands were rejected, those groups had to hide their activities or ways of life. Thus, the demand for privacy is a demand for freedom, the historical basis for the antagonism between personal privacy and public security. Thomas Hobbes describes it as the dilemma of the contract between citizen and state, the tension between personal freedom and the sovereign (state), between security and risk. The sovereign regards privacy as a threat to its security and thus considers it legitimate to collect

---

[78] Richard F. Hixson, *Privacy in a Public Society: Human Rights in Conflict* (Oxford University Press, 1987), 212.
[79] Garfield Benjamin, "Privacy as a Cultural Phenomenon", *Journal of Media Critiques [JMC]* Vol.3(10) (2017), 60.

information about its subjects (citizens and non-citizens alike) to minimize the threat posed by secret thoughts and deeds. This approach received a boost after the September 11 attacks, which created a strong consensus[80] that out there, there is a large and unpredictable terrorist threat, and the way to prevent terrorist activity is by maximizing surveillance and collection of information on individuals and organizations. That was indeed done by the NSA all over the world, as exposed by Edward Snowden.[81]

Amitai Etzioni argues that privacy in the United States is "overprivileged" and that there should be a balance between individual rights – one of which is personal privacy – and public safety and health, which he collectively calls the "common good." Etzioni offers four criteria to help us find the right balance between privacy and the common good: the first criterion is that privacy should only be limited in case of clear evidence of a threat to the common good. Examples discussed in the book are mandatory HIV testing of infants, sexual assaults of children, terrorism and crime;[82] the second criterion is that serious consideration should first be given to whether it is possible to counter the danger without restricting privacy; the third criterion is that the restrictions on privacy should be minimally intrusive; and the fourth is that efforts should be made to minimize undesirable side effects of the privacy-diminishing measures. Etzioni also identifies a pattern that he calls the "privacy paradox": the defense

---

[80] Mark Andrejevic, *Infoglut: How Too Much Information is Changing the Way We Think and Know* (New York: Routledge, 2013), 20.

[81] Stephen Braun, Anne Flaherty, Jack Gillum and Matt Apuzzo (June 15, 2013), "Secret to PRISM Program: Even Bigger Data Seizures", *Associated Press*. Archived from the original on September 10, 2013. Retrieved June 18, 2013.

[82] Ibid., chapters 1-4.

of personal privacy is much weaker when it is violated by the private sector and not by state organizations. Etzioni's approach regards personal privacy as a purely individual interest, making a dichotomy between individuals and society and seeing the struggle between them as a zero-sum game, which means what is good for the individual is bad for society and vice versa. According to Etzioni and others, all we have to do is find the right balance between individual interests and the common good.

However, such an approach diminishes the social importance of personal privacy. According to Daniel J. Solove, it ignores the impact of personal privacy on the nature and creativity of society as a whole.[83] Solove believes that the restriction of personal privacy has implications not only for the affected individual but also for the entire society. As discussed in the previous section, there can be no individual without society, and vice versa. John Dewey puts it this way: "We cannot think of ourselves save as, to some extent, social beings. Hence, we cannot separate the idea of ourselves and our own good from our idea of others and of their good."[84] The question of what came first – individual or society – is irrelevant. Humans are born part of society; they interact with and depend upon each other. Therefore, personal privacy is not a price society must reluctantly pay; it is justified by being good and beneficial for society as well.[85] A society that gives personal privacy to the individuals within it helps them become good citizens who will work for it and contribute to it, making discoveries and inventions that will eventually help the society grow and move forward. As noted by Dewey,

---

[83] Solove, *Understanding Privacy*, 89.

[84] John Dewey, "Ethics (1908)", in: *The Middle Works of John Dewey,* ed. Jo Ann Boydston (Carbondale: Southern Illinois University Press, 1978), 268.

[85] Solove, *Understanding Privacy*, 91.

"Every invention, every improvement in art, technological, military and political, has its genesis in the observation and ingenuity of a particular innovator."[86] Or, in the words of Ralph Waldo Emerson: "Of no use are the men who study to do exactly as was done before, who can never understand that today is a new day. [...] We want men of original perception and original action, who can open their eyes wider than to a nationality – namely, to considerations of benefit to the human race – and can act in the interest of civilization; men of elastic, men of moral mind, who can live in the moment and take a step forward. Columbus was no backward-creeping crab, nor was Martin Luther, [...] nor Thomas Jefferson".[87]

This resonates with the words of Adam Smith in *The Wealth of Nations*[88]: "If each participant in the market of goods seeks to maximize his own benefit, the invisible hand of the market will maximize the benefit of the market as a whole." By analogy, if every individual realizes their full potential, it will bring more benefit to the whole society. However, it would be hard to imagine people who are free and active in the public sphere, realizing their human potential and working for the benefit of society, not having any personal privacy to protect their dignity and their intimate and other relationships.

Privacy, says Solove,[89] has social value; it helps us maintain social relationships and protects us from bullying so we can safely fulfill our autonomy. **Without personal privacy, society cannot exist, and**

---

[86] Dewey, "Ethics (1908)", 207-208.

[87] Ralph Waldo Emerson, "Fortune of the Republic", in: *The Works of Ralph Waldo Emerson in 12 vols, Fireside Edition* (Boston and New York, 1909), Vol. 11 ("Miscellanies").

[88] Adam Smith, *The Wealth of Nations* (London: W. Strahan and T. Cadell, 1776), Book IV, Chapter II, Paragraph 9.

[89] Solove, *Understanding Privacy*, 3.9.

that makes personal privacy a social need as much as an individual one. These two needs are sometimes conflictive and sometimes complementary and must be balanced against each other.

A Metric for Calculating the Extent of Non-Knowledge of Personal Privacy

The metric I propose for calculating the extent of non-knowledge of personal privacy is based on the concept of social distance, described by Georg Simmel as the effect of the information and knowledge people have about each other.[90] According to Claude Shannon's information theory, information is measurable and quantifiable. Here, I present a metric for measuring specific categories of personal privacy in a given situation. The metric is based on the basic components of privacy information, described in the following table:

---

[90] Simmel, "How is Society Possible?", 372-391; Murphy, "Social distance and the veil", 34-35.

Information category	Information elements	Information Index
Private space information	Information related to our private space is defined by social norms and regulations.	1
Body-related information	Body appearance, physical characteristics, physical location, "passport identity" information.	2
Mind-related information	Character traits, mental properties, and preferences (including gender/sexual preferences).	3
Information about actions	Information about our habits, deeds, behavior or routine.	4
Property information	Information about our physical and/or intellectual property.	5
Information about external entities	Information about other entities that are nodes in our network and have information about us.	6
Relationships with external entities	Information about our relationships with other entities: the existence of a relationship, the nature and intensity of the relationship.	7
Autonomy	Information about what can diminish our autonomy, i.e., our ability to be the exclusive decision-makers regarding our own thoughts, feelings, actions, desires, and relationships with others (including relationships with government and corporate entities).	8
Anonymity	Any piece or pieces of information that can – alone or when combined – breach our anonymity, i.e., our ability to remain unidentified in the public sphere.	9
Identity	Information about our various identities.	10

Numeric Calculation of the Extent of Non-Knowledge of Personal Privacy

PI (Personal Information) – All the information you know about yourself.

SI (Social Information) – All the information about you that is known to the public.

Based on these, the extent of non-knowledge of personal information (all the information about you that is unknown to the public) is *EP = PI - SI*. This is the mathematical formulation of social distance.

The social distance of any given individual *i* consists of the following parameters:

PI_i^j is the knowledge individual *i* has about themself regarding information category *j* (e.g., *i*'s knowledge about *i*'s own body). Let its range be $0 \leq PI_i^j \leq 1$. The two extreme cases are $PI_i^j ,= 1$, i.e. individual *i* knows everything about their body (including physical flaws, complete health state and latent genetic features), and $PI_i^j = 0$, i.e., individual *i* knows nothing about their body. Both are very unlikely, though theoretically possible.

WP_i^j is the significance (weight) category *j* has for individual *i*. It is, of course, subjective, varies from one individual to another, and depends on context and social norms; still, it can be assigned an exact value under given circumstances. Its range is $0 \leq WP_i^j \leq 1$.

$SI(l)_i^j$ is the knowledge another individual or entity has about individual *i* regarding information category *j* (e.g., knowledge about individual *i*'s body, including physical flaws, complete health state and latent genetic features). There can be two kinds of "others": a confidant, or a public organization / Internet company. The range of this parameter is also $0 \leq SI(l)_i^j \leq 1$, the extreme cases being $SI(l)_i^j = 1$ (the other has all the information about category j of individual i) and $SI(l)_i^j = 0$ (the other has no information at all about category j of individual i). Both are very unlikely, though theoretically possible. For simplicity's sake, the metric ignores the distinction between the two different kinds of "others".

Thus, based on the above definitions, we can calculate, for example, the extent of non-knowledge of Facebook about individual i using the following formula:

$$EP(i,l) = \sum_{j=1}^{10}(WP_i^j \, (PI_i^j - SI(l)_i^j)) \ (l \text{ representing}$$

Facebook)

Although the metric is well-defined, it may be difficult to use in practice. This difficulty can be overcome by using an N-size representative sample of "others," which would include confidants, Internet companies, etc. N can be relatively small as long as it provides an accurate representation of the public and takes into consideration the extent of knowledge by dominant social entities, such as government organizations or giant corporations like Google, Facebook and Amazon, which make use of people's private information.

The M sample size should be big enough to be a representative sample of individuals with given characteristics. It can be compiled from multiple suitable big data pools. The sampled group's average extent of non-knowledge should be calculated as follows: $DI = \frac{1}{M} \, (\sum_i \sum_l EPi)$ for $i = \{1, 2, \ldots\ldots M\}$, $l = \{1, 2, \ldots\ldots N\}$.

This provides a numeric value that can serve as a metric for measuring information privacy level. The metric can be used by regulatory organs and legal institutions to measure the extent to which

Facebook, for example, meets the GDPR requirements. Here is a numerical example:

Let us assume, for simplicity's sake, that every $WP_i^j = 1$ for every individual i and for every category j of personal privacy. Without loss of generality, I will assume that *SI(1)* represents the knowledge that an Internet company (e.g., Facebook) has about individual i. The following table is an example of the extent of knowledge and non-knowledge:

j index	Information category	*PI* – Personal Information	*SI(1)* – Social information	EP_i^j
1	Private space information	0.7	0.6	0.7-0.6 = 0.1
2	Body-related information	0.9	0.5	0.9-0.5 = 0.4
3	Mind-related information	0.8	0.3	0.8-0.3 = 0.5
4	Information about actions	1	0.7	1-0.7 = 0.3
5	Property information	0.9	0.8	0.9-0.8 = 0.1
6	Information about external entities	0.7	0.7	0.7-0.7 = 0
7	Information about relationships with external entities	0.9	0.6	0.9-0.6 = 0.3
8	Information about decisions (autonomy)	0.8	0.4	0.8-0.4 = 0.4
9	Anonymity information	0.5	0.5	0.5-0.5 = 0
10	Identity information	1	0.8	1-0.8 = 0.2
	Average	0.82	0.59	0.23

Based on these data, we can make a graphic representation of the state of individual i's privacy in relation to the Internet company:

The black circle represents the average personal privacy calculated above.

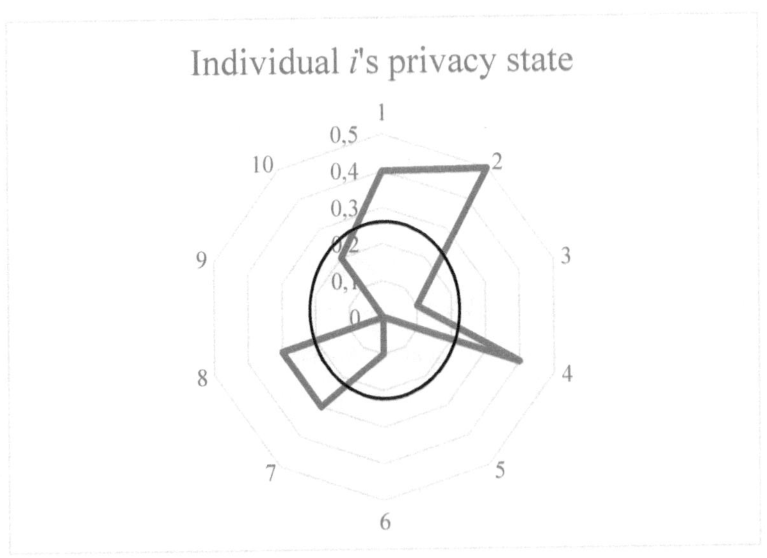

Chapter 3

Personal Privacy as a Cultural

Phenomenon

Humans have been preoccupied with personal privacy in its different aspects since the dawn of civilization. According to the Bible, the first thing Adam and Eve did after tasting from the Tree of Knowledge was to cover themselves to keep their body private.[91] Throughout history, the meaning of "personal privacy" has evolved and changed, but it has always remained a key concept in human interactions. The modern conception of personal privacy, as defined by Western civilization,[92] has its roots in ancient Rome. In Roman times, property owners had the right to personal privacy in regard to their property. This aspect of personal privacy as ownership and control of one's private property and private space still exists today and is present in the discourse of personal privacy; however, in modern times, the concept of personal privacy was extended to also include the mental sphere – thoughts, ideas, feelings and emotions – as described in Ferdinand Schoeman's anthology[93] and in Michael Birnhack's book.[94] This was a result of the growing recognition of the

---

[91] Genesis 3, 6-11.

[92] In this book, the terms "civilization" and "culture" are sometimes used interchangeably.

[93] Schoeman, "Privacy and intimate information", in: *Philosophical Dimensions of Privacy: An Anthology*, 1-34; Solove, *Understanding Privacy*, 174-179.

[94] Birnhack, *Merkhav Prati: Ha-zkhut la-pratiyut...*, 24-35.

rights of the individual as an autonomous being, which increased the need for personal privacy.[95] Consequently, the social definition of personal privacy changed from a physical space to something more abstract.

Culture as a Three-Dimensional Space

I would like to look at culture as a universe that consists of meanings and has three dimensions that together comprise the cultural space, as defined by Ilya Levin: (1) values; (2) legal regulations and social norms (folkways and mores); (3) knowledge.[96] This allows us to represent any given culture, e.g., the Athenian culture in century IV BCE, on a three-dimensional chart where each pair of axes forms a plane that defines a different aspect of the culture: (a) Social culture – a combination of values and regulations; (b) Technological culture – a combination of knowledge and regulations; and (c) Spiritual culture – a combination of knowledge and values.

The model chart looks like this:

---

[95] Arthur Schafer, "Privacy: A Philosophical Overview", in: *Aspects of Privacy Law: essays in honor of John M. Sharp,* ed. Dale Gibson (Toronto: Butterworths, 1980), 2-3.
[96] Ilya Levin, "Cultural trends in a digital society", *Proceedings of TMCE 2014 (19-23 May 2014), Budapest,* eds. I. Horváth and Z. Rusák, 17.

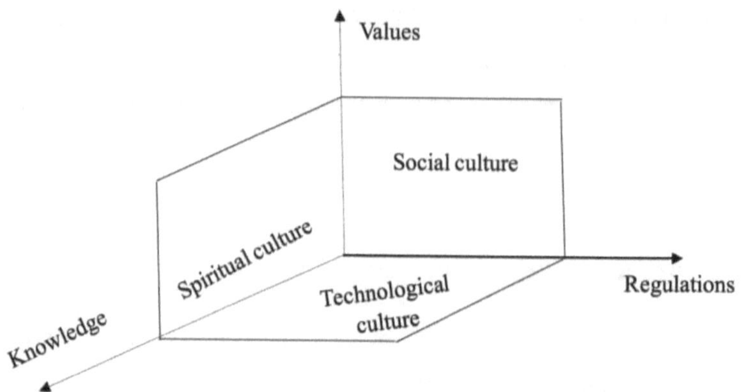

The planes between each two axes form the following aspects of culture:

- The main elements of **spiritual culture**: Art, philosophy, religion (mythology) and beliefs.
- The main elements of **social culture**: Politics,[97] law and ethics.
- The main elements of **technological culture**[98]: Science, technology and engineering.

I will now use this model of culture as a three-dimensional space to analyze privacy in several different societies, which will serve

---

[97] Both in the sense of government and state institutions, and in the sense of international relations and multi-national organizations.

[98] There is a mutual influence between technological culture and privacy. Low technological culture allows limited privacy, as I will show in my analysis of privacy in primitive societies. Advent of technological culture has allowed private physical space, like in ancient Greece and Rome. However, I believe that ever since the emergence of industrial technologies in the 18th century, every consequent technological development has been threatening to reduce privacy. I will discuss it in further detail in Chapter 9 – Personal Privacy in the Age of ICTs, and try to show when and how this turning point took place.

as case studies. An individual and collective claim to privacy exists in all of the societies I examine; however, its manifestations depend on the culture and social norms of each particular society. Privacy is universally applied to three basic elements of human life:[99] **the elements of our identity, our intimate familial relations, and our social relations.** Every society develops social norms aimed to protect privacy in relation to these three basic elements.

Personal Privacy in Primitive Societies

For the purposes of this book, I will define a primitive society as an illiterate society.[100] In such societies, knowledge, art, ideas and other cultural materials are received, retained and passed on to the next generations in oral form.[101] Orally transmitted traditions may include folk tales, songs, ballads, poems and prose. This way, the society passes on its history, literature, laws and all its other cultural knowledge without written documentation. The study of primitive societies began in the 19th century when they were discovered by Western colonialists. Primitive societies were common all over the world – in Africa, Australia, Polynesia, North and South America, and other places. Anthropological research shows many similarities between different primitive societies in the aspect of personal privacy, and therefore, I refer to them here with some degree of generalization. Many prominent anthropologists who

---

[99] Westin, "The origins of modern claims to privacy", 61.

[100] The term "primitive" is not used here in the moral judgmental sense, but in the purely socio-anthropological sense.

[101] Richard A. Posner, "A theory of primitive society, with special reference to law", *The Journal of Law and Economics*, Vol. 23 (1) (April, 1980), 1.

studied primitive societies, e.g., Dorothy Lee,[102] who studied the Tikopia people; Margaret Mead,[103] who studied the Samoan society; and Livingston Jones,[104] who studied the Tlingits of Alaska, claimed those cultures lacked social norms that reflect respect for privacy – at least in its Western sense: the collective (common good) was always prioritized over the individual. People always preferred to be in the company of their fellows, and people who happened to be lonely, childless or widowed would try to change that as soon as they could. Margaret Mead wrote that Samoan houses do not allow any sexual privacy because they have no walls, that even such most intimate moments as death and birth are open for everyone – adults and children alike – to see, and that both adults and children walk around nearly naked at every hour of day or night. Livingston Jones wrote that the Tlingits tend to live in large extended families in private spaces that may be freely entered by anyone and do not allow any privacy by Western standards.

Despite these and other similar findings, Alan Westin believes, based on a close study of findings from 200-300 primitive societies, that they all have three general aspects of privacy:[105] (1) Individual and collective norms; (2) Privacy and solitude; (3) Curiosity and control. Westin bases his arguments on studies conducted by the Department of Anthropology at Yale University.

---

[102] Dorothy A. Lee, *Freedom and Culture* (Englewood Cliffs, N. J.: Prentice-Hall, 1959), 31-32.

[103] Margaret Mead, *Coming of Age in Samoa: A Psychological Study of Primitive Youth for Western Civilization* (New York: New American Library, 1949), 82-85.

[104] Livingston F. Jones, *A study of the Thlingets of Alaska* (New York: Kessinger Publishing, 2004), 58.

[105] Westin, "The origins of modern claims to privacy", 6, 1-60.

Primitive societies provide the universal need for privacy in three basic aspects: **the elements of one's identity, one's intimate familial relations, and one's social relations.**[106] This is done by means of social distance, which, though manifested differently in every society in accordance with its unique culture, exists universally to guarantee personal privacy as it is defined by each society. Thus, parts of one's personal information remain private, and one may choose which information to share and which to keep to oneself. The Tuareg, for example, wear a veil to conceal personal information; the Javans have a social code that dictates which subjects may be discussed freely and which may not; the Balinese maintain social distance by creating a safe, private space in their homes.

Personal Privacy in Primitive Societies According to the Three-Dimensional Space Model

Values: Basic human distinctions between right and wrong, distinction between individual and society, mutual help, and working together to obtain the means of living.

Norms and regulations: In the absolute majority of societies, people maintain bodily privacy (cover their sexual organs) and privacy during sexual relations.

Knowledge: Knowledge is required for survival, and low technology is used to make simple housing, utensils, and weapons.

Spiritual culture: Primitive societies have no writing. Knowledge, art, ideas and other cultural materials are received, retained

---

[106] Ibid.

and passed on to the next generations in oral form and may include folk tales, songs, ballads, poems and prose. Religions are polytheistic and include a belief that we are never alone, even when alone physically, because the gods are all around us, watching and protecting us.

Social culture: Primitive societies have basic values and social norms that are intended to ensure physical and social survival. Privacy is manifested through maintaining social distance, which includes the covering of sexual organs, privacy of intimate relations, codes of behavior and conversation that allow some physical and mental privacy, norms intended to balance between curiosity and control, and norms that dictate which topics may be discussed in public, what may be shared with others, and what should only be discussed in the private family circle.

Technological culture: It would be hard to talk about technological culture in the sense of science, technology, and engineering, but primitive societies use technology to achieve some privacy. One example is clothes, which are used to cover sexual organs or, as with the Tuareg veil, to create social distance and signify social status. Another is private residences, which, though many times physically open, nevertheless define a private physical space admissible only to a particular person or family.

Let us represent the state of personal privacy in primitive societies on the three-dimensional space chart:

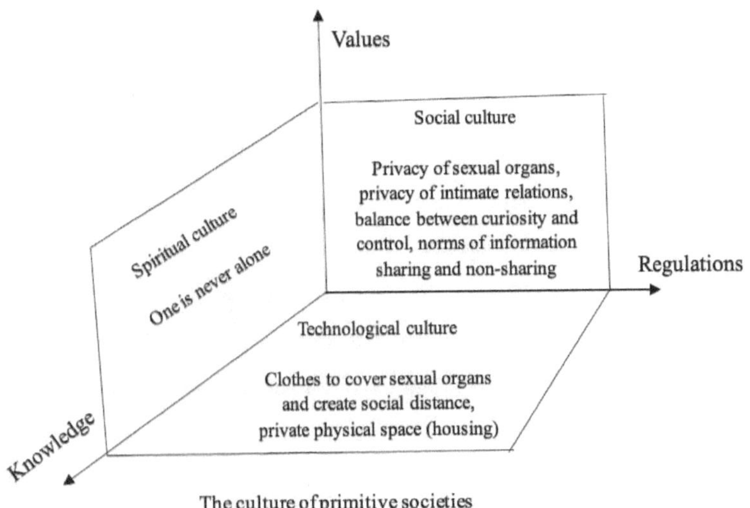

The culture of primitive societies

Personal Privacy in Ancient Athens

Our discussion of personal privacy in Athens will refer to the time between the end of the Peloponnesian War (404 BCE) and the Battle of Chaeronea, in which Athens was defeated by Philip II of Macedonia (338 BCE).[107] It was the time of Plato, Aristotle, and Demosthenes, the golden age of Athenian democracy, and a time that represented the concept of privacy as it was understood and applied in ancient Athens.[108] In order to understand the meaning of privacy in Athens during the period of time discussed, we should first understand what was considered private and what was considered public. Let us begin by explaining the citizens' obligations to their polis (city-state) and the obligations of the polis to its citizens.

---

[107] Barrington Moore, *Privacy: Studies in Social and Cultural History* (Armonk, N. Y.: M. E. Sharpe, 1984), Chapter 2.
[108] Ibid., 81.

Athenian citizens had many different duties. One of them was military conscription, which is a good example of the tension between private and public.[109] The mandatory military service lasted about three years, during which the conscript soldier had to give up most of his private life and go wherever he was required. In return, he was exempted from tax duty and not subjected to the civil legal system. Adult citizens were also forced to finance the Athenian warship fleet but refused to cover all the costs. This is another example of the clash between the individual's personal interest and the public interest of the polis.

The idea of private life as an alternative to public life is mentioned in Greek literature from the 5th century BCE in the narrow sense of non-participation in the political life of the polis.[110] A private life in the sense of autonomy to control one's own everyday life was a luxury only the rich could afford; however, for peasants, it meant the possibility to farm their land in peace.[111] Tradition has it that the Athenian statesman Solon passed a law that allowed citizens who evaded their civil obligations, particularly their military duties, to be deprived of their citizenship. The purpose of the law was to fight people's tendency to choose private life over civic service, and it proves that in Athens, private life was a legitimate option as long as one fulfilled one's obligations to the polis, i.e., maintained propriety in his public life.

Famous Greek philosophers such as Plato, Aristotle, Socrates, Demosthenes, and Xenophon (centuries V-IV BCE) believed that life devoted to philosophical learning was a worthy alternative to public (civic) life. According to Plato's *Apology of Socrates*, Socrates had paid

---

[109] Ibid., 86-90.

[110] Ibid., 117, 133.

[111] Without being intruded by foreign armed forces or required to go to war.

with his life to pursue the ideal of free inquiry,[112] a recognition of the individual's autonomy of independent thought.

Plato sought the ideal social order and regarded privacy as something negative people use to evade their civil duties. In *Republic,* he writes: "the governors [...] will take their soldiers and place them in houses [...] which are common to all, and contain nothing private, or individual".[113] In fact, his views on privacy were so radical[114] that he considered it a threat to good society, declaring, "The state comes first."[115]

Unlike Plato, Aristotle had no firm opinion regarding privacy, but he did acknowledge the inevitable conflict between individuals and society.[116] Like Socrates, he realized that the intellectual pursuit of human happiness requires full freedom and autonomy of thought.[117] Xenophon's position was somewhere in between: he allowed the possibility of private life as part of a broader existence but was against dedicating oneself to private life alone. Unlike Plato, he did allow for privacy in domestic and family life.

Bodily Functions, Family Life and Public Life in Athens

In ancient Athens, the distinction between private and public meant the distinction between polis norms and the norms of behavior in the family

---

[112] Plato, *The Apology of Socrates,* 31D-32A.

[113] Plato, *Republic* (trans. Benjamin Howett), 8.543B.

[114] Plato, *Laws,* 5.738D-E.

[115] Ibid., 5.739B-E. Later in the same section, Plato argues that all property should be common, even what was considered private property.

[116] Moore, *Privacy: Studies in Social and Cultural History,* 125-126.

[117] According to Aristotle, achieving happiness is man's supreme ethical goal.

circle. The public and political spheres were controlled entirely by men and were inaccessible to women; however, private spaces were controlled by men and women together, often mostly by women. Both Plato and Aristotle believed political control must also be extended to the sphere of home and family. Plato, the more radical of the two, said the purpose of the family was to serve the polis, so it should be supervised by the polis to make sure it fulfills its duty – childbirth and rearing to continue the bloodline and preserve the legacy. In Athens, family was considered an element of the polis, and therefore, commitment to the polis preceded commitment to one's family.

One's home was considered one's private space, which may not be entered without the owner's permission. The private space helped maintain the privacy of the female family members from prying eyes. The privacy of each individual member of the extended household – which also included the slaves – was not observed within the private space; however, toileting and sexual relations were to be had in private. The cultural attitude toward nudity was ambivalent: body exposure of adult people in public was unacceptable; however, athletes competed in the nude.

To conclude, in ancient Athens, there was a clear distinction between the public sphere and the private domestic sphere. The public interest usually prevailed over private interest. Individuals had commitments to the polis, which preceded their commitments to their families. Within the domestic sphere, one was almost always exposed to other members of the household. Women, slaves, and low-class residents could not enjoy most of the individual rights of Athenian citizens.

Morality in Private and Public Life in Athens

The Athenians believed that the moral code of behavior should be equally applied in the public and private spheres (particularly in the sphere of family life). Demosthenes said a man should be judged based on his private life as well as his public activity.[118] Plato, too, made no distinction between behavior in the public and private spheres. Aristotle, on the other hand, thought it possible to separate one's moral behavior in the public sphere from one's behavior in his private space.[119] In modern times, we try to completely separate private life from the public sphere.

Despite Demosthenes', Aristotle's and Plato's philosophy, Athenians did not pry into the private lives of candidates for public office and only made sure they had no history of violation of civil rights. In Athens, in contrast to modern-day Western law, one could be put on trial because of one's reputation without having to prove any particular illegal deed.

Personal Privacy in Ancient Athens According to the Three-Dimensional Space Model

Values: The individual is a free citizen.[120] The polis comes first. The individual has limited personal autonomy and is expected to behave the same way in public and private life.

Norms and regulations: Civil rights, legal system, rights of property, civil duties (military conscription, taxes), polis interference in

---

[118] Demosthenes, *De Corona,* 8.
[119] Moore, *Privacy: Studies in Social and Cultural History,* 156.
[120] This discussion refers only to the citizen class.

family life (particularly in caring for widows and orphans).[121] Limited bodily privacy, the privacy of intimate relations, and friendships are accepted as private personal matters. Private property may include physical assets such as houses or land, as well as wives and slaves.

Knowledge: Mathematics and engineering (Euclides, Pythagoras and their disciples), engineering knowledge that allows the construction of large buildings and ships, astronomy (later known as Ptolemaic astronomy), and marine navigation.[122]

Spiritual culture: Philosophy that explores the essence of nature and humanity, theater that performs plays focused on human conflicts, sculpture. A polytheistic religion with gods who constantly interfere in the human world, often using humans to serve their conflicting interests, and against whom man sometimes rebels.[123] One has no privacy and limited autonomy from the gods, who know all one's deeds, thoughts and feelings, interfere with them and force their will. Well-developed poetry with constructed narratives, which also reveal heroes' weaknesses. One is responsible for one's moral choices (tragedies are the result of making wrong moral choices).[124]

In relation to privacy: Limited autonomy from the gods, responsibility for one's moral choices, and recognition of the individual's autonomy of independent thought.

Social culture: A democratic political regime in which the ecclesia (the assembly of citizens) is the main governing body. Laws and

---

[121] The trial of Socrates is a good example.

[122] Sabetai Unguru, *Mavo le-toldot ha-matematika* [An Introduction to the History of Mathematics] (in Hebrew) (Tel Aviv: Ministry of Defense Publications, 1989), Part I.

[123] E.g., the story of Prometheus.

[124] E.g., *The Iliad* and *The Odyssey*.

court systems are used to manage interpersonal conflicts between individuals and states. Privacy is limited to the domestic sphere, with no privacy within the extended household. The family's autonomy is limited and does not include care for widows and orphans, which is the responsibility of the polis. Every citizen has to fulfill his civil obligations, and in case of a clash between civil duties and family duties, the polis comes first. The polis protects the weaker citizens. There are rights to private property, which may be inherited. One's private space is prohibited for strangers and those who are not part of the extended household. Ambivalent attitude toward nudity: though it was not normative for adults to be naked in public, athletes did compete in the nude. A small society in which everybody knows everybody, and anonymity is neither normative nor desirable.[125] One is expected to behave the same way in public and private life; having "multiple faces" is considered immoral.

In relation to privacy, The public sphere is stronger than the private sphere. Personal privacy is perceived as non-participation in public life. Mandatory tax payment, military conscription, and partial autonomy in caring for the weak. Private space with limited access to strangers, rights of private property, bodily privacy depending on circumstances, and privacy of sexual relations. No anonymity. The same behavior is expected from one in private and in public.

Technological culture: Construction of large buildings and ships that sail the Mediterranean, engineering based on mathematics and physics (Archimedes). In ancient Athens, science and technology were not used to increase privacy.

---

[125] According to Plato.

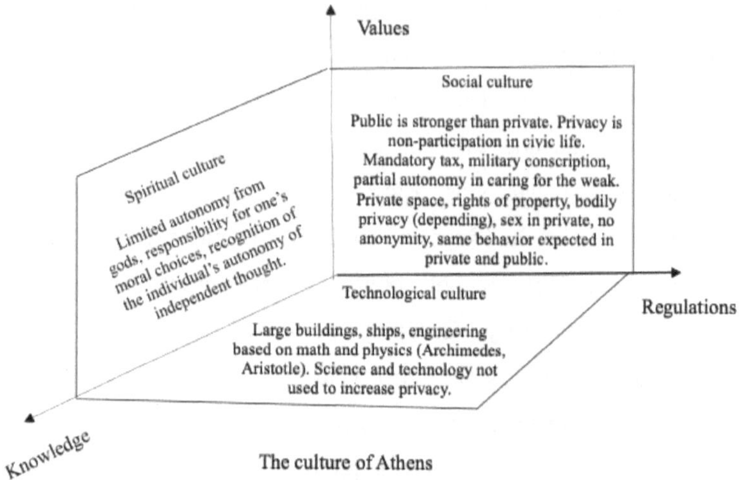

The culture of Athens

Personal Privacy in Rome

Roman culture was, in many ways, a continuation of Greek culture. A very good demonstration of the distinction between the private and public spheres is the story of the struggle between Cicero and Clodius over the fate of Cicero's former house on the Palatine hill. In his speech before the citizens of Rome, Cicero describes the officeholder as a man who has public visibility and is acting in the best interest of the public, as opposed to the private man who puts his personal interest first.[126] He makes a distinction between a public cult, which concerns the entire community, and a private cult – by a private individual or group of individuals. Both Cicero and Clodius understand that one's house is

---

[126] Elisabeth Begemann, "Ista tua pulchra libertas: The Construction of a Private Cult of Liberty on the Palatine", in: *Public and Private in Ancient Mediterranean Law and Religion*, ed. Clifford Ando and Jörg Rüpke (Berlin/Munich/Boston: De Gruyter, 2015), 76.

one's private space, which signifies one's power, and taking it away means weakening one and evoking one's privacy. Both are struggling for the narrative: Cicero tries to portray Clodius as a tyrant and betrayer of public interest, while Clodius paints Cicero as an insurgent who puts his own private interest before that of the republic. Confiscation of houses and land was a common form of punishment in Rome. Cicero's confiscated property was turned into a shrine of the goddess Libertas, but he argues that Clodius had built the shrine for his own use, and thus, it does not serve the public interest, only the private interest of Clodius himself. Thus, Cicero presents the conflict as a struggle between the private interests of two individuals rather than a struggle between his private interest and the public one and uses this to try and reclaim his property. In Rome, the land once dedicated to gods belonged forever after to the public. Shrines and places of worship, even if established and paid for by individuals on their private land, were henceforth considered public property. Speaking more generally, a site's being public or private was determined based on its purpose and use, not on the identity of its dwellers or its financing source. There was a clear distinction between public activities, which took place in public places (e.g., places of worship), and private activities, which took place on one's private property; any given site could only be used one way or another, not both.[127]

In familial matters, there was no such clear distinction between private and public.[128] For Romans, the family was the basic unit of

---

[127] Ibid., 105.

[128] Judith Evans Grubbs, "Making the Private Public: Illegitimacy and Incest in Roman Law", in: *Public and Private in Ancient Mediterranean Law and Religion*, eds. Clifford Ando and Jörg Rüpke (Berlin/Munich/Boston: De Gruyter, 2015), 115.

society. The term "family" referred to the extended household, which included one's wife, children (adult and underage alike) and slaves. The head of the family had absolute control over the other members of his family, including "the right of life and death." Men who put the interest of the republic before their family and private needs were highly praised. Around the time of Emperor Augustus (century I BCE - I CE), the state began to interfere in family life by penalizing adulterous women and married couples who did not have children. Thus, sexual relations ceased to be one's private business and became the object of public supervision and legislation. Roman law also defined the status of illegitimate children and children born of an incestuous relationship[129]. Such children did not merely belong to the private sphere; they were a matter of public concern and public laws and mores.

Personal Privacy in Ancient Rome According to the Three-Dimensional Space Model

Values: The individual is an autonomous, free citizen. The republic is the sovereign, but its power over free citizens is limited. One has moral responsibility for one's actions.

Norms and regulations: A distinction between private and public spaces. The private space is one's private property protected by law; places of worship are public spaces. A distinction between a public officeholder and a private individual and between public and private

---

[129] The legal and normative details are beyond the scope and purpose of this book. The only reason I mention it here is to emphasize that some familial relationships were not considered private matters, and the state could legally interfere in them.

activities. The power of the ruler is limited.[130] The family is the property of the head of the family. There are laws concerning sex: some forms of sexual relations are forbidden by law, meaning the state may interfere in those matters.

Knowledge: Knowledge of mathematics and engineering is roughly the same as that of Athens.

Spiritual culture: The Roman religion, mythology and art were a continuation of the Greek. The Romans developed the republican regime and law.

Social culture: A space was deemed private or public based on its use, not on its ownership or location (a public site may stand on private land). Despite this distinction, the homes of Roman aristocrats often served as public as well as private places. An officeholder is defined as a man acting in the interest of the public. The law protects individuals' private property from being overtaken by the ruler or by other individuals. Those who put the interest of the republic before their family and private needs are praised.[131] Legally, children born of incest or close-kin marriage do not belong only to the private, familial sphere; they are a matter of public concern and public laws and mores, and thus are in some way public.

In relation to privacy, Even though there is a distinction between the public and private spheres, it is not absolute. Private property rights are protected by law. The body is supposed to be covered. No full autonomy of sexual relations. Sex is had in the private space. The sovereign has limited authority to interfere in people's private affairs.

---

[130] When a ruler exceeds his power, he becomes a tyrant, i.e., an illegitimate ruler.

[131] Evans Grubbs, "Making the Private Public", 116.

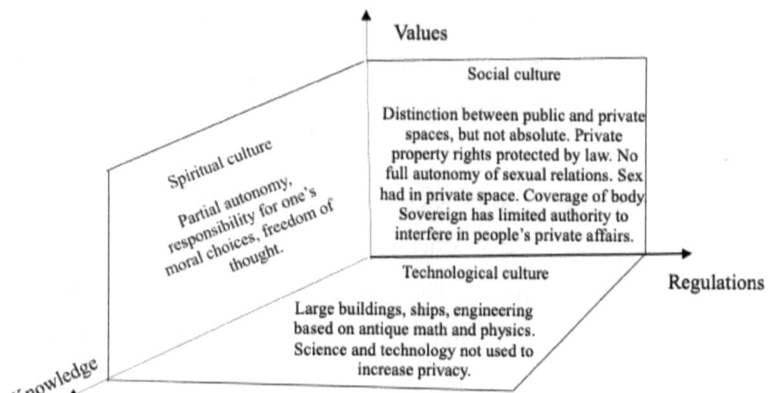

Values
Regulations
Knowledge
Spiritual culture
Partial autonomy, responsibility for one's moral choices, freedom of thought.
Social culture
Distinction between public and private spaces, but not absolute. Private property rights protected by law. No full autonomy of sexual relations. Sex had in private space. Coverage of body. Sovereign has limited authority to interfere in people's private affairs.
Technological culture
Large buildings, ships, engineering based on antique math and physics. Science and technology not used to increase privacy.

Chapter 4

Privacy in the Modern Era

The Roots of the Modern Conception of Privacy

The modern-day conception of personal privacy developed out of the humanistic values of the Age of Enlightenment. It was the age of the birth of Western democracies, which focused on people as individuals who have rights, including the right to personal privacy. While in the past, privacy[132] meant withdrawal from society, or at least withdrawal from political life and public affairs, in modern times, it was extended to also include the mental sphere, i.e., thoughts, ideas, feelings and emotions, as a consequence of the growing recognition of the rights of the individual as an autonomous being.[133] As a result, the social definition of personal privacy changed from a physical space to something more abstract. The first signs of this change can be seen in an 1890 article called "The Right to Privacy,"[134] in which authors Samuel Warren and Louis Brandeis stand for the individual's right to be protected from external intrusion and discussion of his affairs. That became the modern Western interpretation

---

[132] Thomas W. Laqueur, Naomi Fry, "Ha'im pratiyut hi modernit? 'Ha-khet ha-prati' shel ha-onenut" [What is Modern about Privacy? What is Private about the 'Private Vice'?], *Zmanim: A Historical Quarterly*, 104 (Fall 2008), 12-17 (in Hebrew).

[133] Schafer, "Privacy: A Philosophical Overview", 2-4.

[134] Cayce Myers, Samuel Warren and Louis Brandeis, "The Right to Privacy", 4 Harv. L. Rev 193 (1890), *Communication Law and Policy*, Vol.25 (4) (October 2020), 519-522.

of privacy. It began with the emergence of the concepts of self and selfhood, for there can be no personal privacy without an individual who is a distinct, autonomous subject with a right and ability to claim their own rights, including the right to privacy.[135] According to Dror Wahrman, the conception that formed during the late 18th and early 19th century was that one's inner self, one's core identity, was fixed and unchangeable: there could be no gender swaps or multiple identities. A law passed in France soon after Robespierre's execution prohibited name-changing.[136] This conception of a fixed, "atomic" subject stood in contrast to the older conception of a gap between one's inner self and one's external presentation, which may change depending on the circumstances of one's life.

Individualism: A Concept Born from Enlightenment

Perhaps for the first time in history, people in Western societies had an opportunity for self-realization, which led to an expectation for fulfillment of the entire human potential. This strife for self-fulfillment is a major human experience in the modern age.[137] The idea of individualism necessarily suggests personal privacy, and so the rise of individualism allowed personal privacy to flourish in modern Western societies. Individualism has been discussed by some of the greatest

---

[135] Dror Wahrman, Yaniv Farkash, "Im ein ani li: Ha-prat ha-naor be-idan ha-mahapekhot" [The Enlightenment and Selfhood: A New Perspective on the Age of Revolutions], *Zmanim: A Historical Quarterly*, 104 (Fall 2008), 62-71 (in Hebrew).

[136] Ibid.

[137] Lior Rabi, *Omes ha-individualiyut – ha-shorashim shel ideal ha-individualiyut ha-moderni* [The Burden of Individuality: The Origins of the Modern Ideal of Individuality] (in Hebrew) (Haifa: Pardes, 2009), 7.

philosophers of the Age of Enlightenment. **Immanuel Kant** believed that every human being – not just members of the social elites – possesses autonomous rational abilities that are universal,[138] i.e., that some rational abilities are shared by all human beings. The emphasis in Kantian individualism is on what is common to all human beings, not on what is different and unique to each individual.[139]

Karl Marx believed he had managed to uncover the laws of history.[140] The purpose of human history, he believed, was to create an economically equal, socially just, class-free and homogenous society inhabited by impersonal proletarian individuals. According to Marx, contemporary individualism was unnatural, product of a capitalistic society that sanctifies the rights of private property and personal freedom to preserve the power relations between the classes.[141] After the proletarian revolution, every individual would have the same impersonal class consciousness of moral mission. Marx rejected the idea of different, unique individuals and believed that as part of the class struggle, the working person must shed every personal characteristic and personal interest to become a new type of individual – the "general proletarian," driven by class needs. That, according to Marx, was the fulfillment of our historical purpose, regardless of our individual traits.[142]

Friedrich Nietzsche put individualism before all the other ethical requirements.[143] He believed one must realize one's full potential but agreed that not all people would be able to reach that goal. Nietzsche

---

138 Ibid., 42.
139 Ibid., 8.
140 Ibid., 157-158.
141 Ibid., 181-182.
142 Ibid., 13.
143 Ibid., 188-189.

did not believe in the possibility of an individualism-based society and, therefore, in the possibility of a liberal democracy that would ensure individualism for all. According to Nietzsche, every person is different, and it is important to nurture each person's individuality and put the interest of the individual before the public interest. He believed one must, first of all, be true to oneself and one's feelings and regarded conformity as a weakness.

Jean-Jacques Rousseau believed that the natural state of man before he became bound by social constraints, was one of freedom and equality, which everyone could enjoy because nature's resources were not owned by anyone.[144] Social life has made man a slave to property and to the desires and whims of others. As a way to settle the conflict between individual and society, Rousseau introduced the concept of the "general will" as opposed to the particular (individual) will and said that "the particular will tends, by its very nature, to partiality, while the general will tends to equality,"[145] and therefore, subordinating the particular will to the general will would ensure equality between people. Rousseau's approach acknowledges the inherent conflict between individual and society: on the one hand, society gives the individual safety and security in their broadest sense – the laws and social norms protect the individual's physical safety from violence and bullying, while a liberal democratic society that acknowledges individual rights also defends the individual's freedom to realize them; but on the other hand, there is an inevitable clash between the needs of the individual and the needs of society or other individuals because every individual has

---

[144] I disagree that man can exist without society. Humanity and society were born at the same time, and neither has existence without the other.
[145] Jean-Jacques Rousseau, *The Social Contract* (trans. G. D. H. Cole), Book II, Chapter I (Amsterdam, 1762).

different personal and social needs which may conflict with those of others.

According to **Baruch Spinoza**, free will is not arbitrary but necessary and bound to the laws of nature because it comes from within us. The more developed one is, the more one is aware of the laws of nature and the fact that they stem from the laws of God. That awareness is freedom, and it is threatened by anything that is forced from the outside and does not stem from our inner nature.[146] Spinoza's concept of freedom does not mean arbitrary and unrestrained free will but an increasing awareness of "the necessary freedom" until we become united with God. That union can be interpreted as the dissipation of the individual as a separate being and, consequently, as the dissipation of privacy. Spinoza's solution to the conflict between the public and private interests is blending into the universal good out of necessary freedom.

Individualism and Its Internal Contradiction

Individualism as a philosophical concept contains an internal contradiction that stems from the interpretation of the concept of human freedom. Seemingly, human freedom means absolute freedom to realize one's individuality without any constraints. However, this would be impossible in a human society, where individuals must give up some of their personal freedom to be able to live together and cooperate. Freedom can never be absolute and unconstrained; it must be limited, at least to the extent suggested by John Stuart Mill. In other words, there must be some restriction on the autonomy of action, which is a key element of

---

[146] Ibid., 40.

personal privacy.[147] Thus, even though personal autonomy is a precondition for self-fulfillment, it (and with it – personal freedom) must be restricted a priori out of social necessity.

All the prominent Enlightenment philosophers reviewed in the previous section seem to have been aware of this internal contradiction and offered different ways to solve it by integrating the individual into something greater than themself. The a priori assumption of the liberalist approach born out of the American and French revolutions was that the inherent conflict between individuals and society was inevitable and unsolvable, and therefore, to allow people to fulfill their individuality, there was a need for an appropriate system of social and political balances. But even then, self-fulfillment would be limited, leading to a blur between the universal and the individual.

Personal Privacy as a Derivative of Individualism

Individualism is the acknowledgment that every person is an individual in their own right. This acknowledgment gave birth to the liberal democratic approach. Modern-day democracy is founded on the values of the French Revolution and the American War of Independence: liberty, equality and justice. According to Allen Wood, "liberty" in the liberal tradition usually refers to individuals' privacy,[148] the sphere where we can do whatever we want without interference from other individuals or the state. For a liberal, liberty is the claim to well-defined and agreed borders of the private sphere, **the private sphere being the area in which one can and must fulfill one's individuality**. Thus, the

---

[147] Inter alia, because it allows freedom of intimate relations.

[148] Allen Wood, *Hegel's Ethical Thought* (Cambridge: Cambridge University Press, 1990), 36.

discussion of personal privacy becomes a discussion of the boundaries of the private sphere. Even if there is a general consensus in liberal thought that there must be a line between private and public, the placement of that line is culture- and time-dependent, and thus frequently changing.

According to the liberal democratic approach, every person is a unique individual who has the right and duty to realize their full potential.[149] Liberalism acknowledges the conflict between individual and society and regards the combination of democracy and government as a solution. The government provides the social frame that allows people to be part of something greater than themselves, while democracy provides the opportunity for political self-expression, as it assumes every individual is rational and autonomous. Participation in political life is a manifestation of human reason and autonomy and would not be possible without democracy. Unlike the previously reviewed philosophies, liberalism regards the conflict between individuals and society as a given state of events that cannot be resolved and as a constructive conflict that helps both individuals and society grow. It allows the individual to express themself freely without having to hide or be persecuted for their beliefs, thoughts or actions, thus helping protect minority rights and prevent tyranny of the majority. From the social aspect, it helps build a society that is based on tolerance, acceptance of the other, and pluralism of opinions and ideas.

Personal freedom is the right to do anything as long as it does no harm to others. In the words of John Stuart Mill: "The individual is not accountable to society for his actions, in so far as these concern the

---

[149] Rabi, *Omes ha-individualiyut*, 59-60.

interests of no person but himself."[150] This definition, which lies at the base of the modern conception of individuality and personal autonomy, is not as clear as it may seem. There are many situations in which it is hard to draw a clear line between actions that only concern ourselves and the implications those actions may have on others. Smoking in public places, for example, had been considered people's private business for years until the harm of passive smoking became known. R.W. Emerson (and other American philosophers) called America "a nation of individuals" and said that "the less government we have, the better," which means that in order to allow people to reach their full potential, government intervention should be minimal.[151] This universal idea has united America as a country of its citizens rather than a nation-state. Rejecting the idea of interfering in other people's lives is also used to justify social gaps and to defend private property. Society has no commitment to its members, and poverty is considered the poor's own fault: supposedly, everybody has an equal opportunity to have a good life; those who have failed to take it are to blame for their own condition, and society has no responsibility for that.[152]

According to the liberalist approach, our mission in life is to realize our unique individuality. "Liberty" comes to mean the freedom to be whatever you make of yourself.[153] The state's job is to provide the

---

[150] John Stuart Mill, *On Liberty* (London: J. W. Parker and Son, 1859), Chapter V, 178.

[151] Or, as Hanry David Thoreau quoted Thomas Jefferson, "I heartily accept the motto, – 'That government is best which governs least'" (Hanry David Thoreau, *Civil Disobedience*).

[152] Rabi, *Omes ha-individualiyut*, 256-257.

[153] Amos Harpaz, *Sheker Ha-Individualism: Spinoza, Hegel, yeha-dimui ha-kozev shel ha-'ani' ha-moderni* [Falsity of Individualism: Spinoza, Hegel, and the False Image of Modern Man] (in Hebrew) (Tel Aviv:

social and political frame that would allow you self-discovery and self-development. "Equality" in liberalism means that all people are born equal. All are equal before the law, which is the same for everybody, and equal politically, i.e., each person has one voice. This is a manifestation of the Kantian idea that every person naturally possesses an "autonomous rational," from which stem the values of political equality and freedom to make autonomous rational decisions – the philosophical rationale behind secret elections. "Justice" is seen mainly as the provision of equal opportunity for everybody to fulfill themselves. Generally speaking, the democratic idea was originally designed by various philosophers of the Enlightenment as a way to balance the needs of the individual against the needs of society.

The three principles of liberalism – liberty, equality and justice – together gave birth to the modern conception of personal privacy. Liberty requires a private sphere where one can fulfill one's individuality; equality is the acknowledgment that all people possess the rational ability to set universal rules of ethics which would define the limits of individual autonomy and the social norms that help distinguish between private and public; justice defends the natural and universal human right to self-fulfillment. Out of these three principles, designed by thinkers who had been fully aware of the seemingly insoluble tension between individual and society, the unique and the universal, was born the personal privacy of modern Western democracies, which sharply contrasts with Marx's vision described earlier in this chapter. Dror Wahrman puts it well: "The concept of selfhood is at the base of what characterizes the modern era [...] privacy follows the same logic: one cannot talk about "private" or "right to privacy" without assuming the

---

Resling, 2013), 40.

existence of a clearly distinguished autonomous subject contained within itself."[154]

The function of privacy is to resolve the tension between individual and collective by creating a private sphere in which one is entitled to fully realize one's self-potential. Beyond it lies the public sphere, where one must abide and be bound by the rules and the social norms. The line between these two spheres is culture-dependent and often blurred.

The belief that every person has a unique individuality intensified after WWII. In fact, individuality had become so sacred that its realization became a must.[155] Even though every individual is made up of the same universal building blocks, such as common human physiology, genetics and cognitive abilities, the particular combination of those, together with every person's unique history, is what creates each individual's unique selfhood.[156] It is the individual's duty to realize their individuality and society's duty to allow them to do that. The perception of a human being as a unique individual who has liberties and rights is summed up in the sentence: "The individual is not accountable to society for his actions, in so far as these concern the interests of no person but himself."[157] Distinguishing between those actions that concern nobody but ourselves and those that do concern other people is not that simple because there are actions for which we are accountable to others, so-

---

[154] Wahrman and Farkash, "Im ein ani li", 62-71.

[155] Rabi, *Omes ha-individualiyut*, 255.

[156] Jorge J. E. Garcia, *Individuality: An Essay on the Foundation of Metaphysics* (New York: State University of New York Press, 1988), 234.
This is similar to the way every human's DNA is made up of the same four basic units, but each person's complete DNA is unique.

[157] Mill, *On Liberty*, Chapter V, 178.

called "confidants" – individuals or organizations to whom we disclose parts of our personal privacy. Confidants include close friends and family members as well as healthcare providers, regulatory agencies such as tax authorities, etc.[158]

To conclude, the modern Western conception of personal privacy was born from the individualist approach that developed in the Western democratic culture. Personal privacy cannot exist without individuality because it is an attribute of individuals; without individuality, any discussion of privacy would be meaningless.

---

[158] As we will later see, individuals give accounts of their actions whether they want to or not. Moreover, such accounts can be obtained from external data pools. I will also elaborate on the definition of "confidants" and the extent of their commitment to maintaining our personal privacy.

Chapter 5

The Industrial Revolution and the New Definition of Personal Privacy

Urbanization Redefines the Private and Public Spheres

The Industrial Revolution desperately needed strong working hands to personalize the assembly lines. So, millions of young people moved from rural areas to big cities, creating a new, industrial society.[159] The process of urbanization brought together large numbers of strangers who previously lived in small rural communities where everybody knew each other. The family structure changed, too: large multigenerational households sharing the same rural space gave way to nuclear units of father, mother and a few children.[160] The extended family could not move to the city together, whereas singles and smaller nuclear families could make the necessary move in search of new jobs. That was a tremendous social revolution, which sociologist Ferdinand Tonnies

---

[159] According to Wikipedia, only 3% of the world's population lived in cities in 1800, compared to 15% at the turn of the 20th century. By 2007, more than 50% of the world population were living in cities, and it is predicted that by 2050 about 64% of the developing world and 86% of the developed world will be urbanized. https://en.wikipedia.org/wiki/Urbanization, https://en.wikipedia.org/wiki/Industrial_Revolution. (Accessed March 22, 2024).

[160] Toffler, *The Third Wave*, 44-45.

defined as the transition from community to society.[161] In the city, one became a lone individual in a society made up of strangers, alienated from most of one's surroundings. New social norms and definitions concerning interaction between those strangers began to form in order to allow cooperation, without which there can be no society. According to Georg Simmel, in the city, every man was for himself, and that increased personal privacy – mainly in the sense of greater anonymity and loneliness, but also in the sense of greater autonomy and personal responsibility – in other words, greater individuality.[162] This social transformation also triggered the formation of new laws and social rules that would make a society made up of anonymous strangers possible.

Thus, urban society is a society of strangers in which intimate and personal relationships are limited to the nuclear family, while anyone outside the family unit is a stranger. This is the very essence of anonymity: a stranger among strangers has no individuality, nothing to distinguish them from the other strangers surrounding them. Every individual is anonymous to all the other faceless individuals around them. There is a clear distinction between interactions in the private sphere – the nuclear family, where every individual is unique and has a unique relationship with every other family member – and the public sphere, where everyone is anonymous and has no individuality. Urbanization defined the private sphere as the space where the nuclear

---

[161] A community is a small social unit in which everybody knows each other, whereas a society is a collection of many people, most of whom are strangers to one another. Ferdinand Tonnies and Charles P. Loomis, *Community & Society* (East Lansing: Michigan University Press, 1957), 19-20.

[162] Simmel, "How is Society Possible?".

family resides, and the relationships within it shape and evolve. Anything beyond it is considered the public sphere.[163]

Industrial Society Shapes the Modern-Day Public Sphere

The system of mass production – also called "Fordism" after its pioneer, Henry Ford[164] – reached its peak in the late 19th and early 20th century. According to Bob Jessop, there are four levels of Fordism:

Capitalist labor process: Large masses of generic unskilled workers employed along   Taylorist lines,[165] each performing one or a few tasks that require no special skill or expertise. The produced goods are standardized – assuming the consumers are also standard and generic

---

[163] In the next chapters, we will see how these definitions are being challenged in the age of ICTs.

[164] Bob Jessop, "Fordism and Post-Fordism: A Critical Reformulation", in: *Pathways to Industrialization and Regional Development*, eds. Michael Storper and Allen J. Scott (London: Taylor & Francis, 1992), 1-41.

[165] Taylorism: Production efficiency methodology that breaks every action, job, or task into small and simple segments which can be easily analyzed and taught. Introduced in the early 20th century, Taylorism (1) aims to achieve maximum job fragmentation to minimize skill requirement and job learning time, (2) separates execution of work from work-planning, (3) separates direct labor from indirect labor, (4) replaces rule of thumb productivity estimates with precise measurements, (5) introduces time and motion study for optimum job performance, cost accounting, tool and work station design, and (6) makes possible payment-by-result method of wage determination. https://indianmoney.com/financial-dictionary/t/taylorism (Accessed January 12, 2021)

– except for some general adjustments such as clothes size and trends. Mass production does not target any specific individual customer but a mass market of generic and anonymous consumers.

Accumulation regime: A mass of people and jobs accumulated in the same physical space – i.e., large numbers of workers, assembly lines and industries concentrated and accumulated in the industrial cities. The manufacturing companies usually control all of the production processes, resources and materials. Involves a "virtuous circle of growth": by increasing productivity, wages rise, resulting in higher productivity, demand, investment, and operational efficacy.

Social mode of economic regulation: In return for the monotonous work and constant demands to increase productivity, the workers are promised continuous payment raises that would improve their quality of life, increasing their purchasing power and allowing them to buy the products they make. This led, on the one hand, to the establishment of giant corporations (some of them monopolies) and, on the other hand, to the formation of trade unions whose job is to protect the interests of the assembly workers and make sure the promises of continuous financial rewarding are kept. The third side of this triangle is the government, whose task is to find and maintain a balance between the conflicting interests of the capitalists and the workers.[166]

Generic mode of "societalization": The vast majority of the population depends on an individual wage and/or social benefits provided by the state to satisfy their consumption needs. As a mirror of the system of mass production, there is growing mass consumption of standardized, mass-produced commodities. Mass production and mass consumption complement and stimulate each other. Mass production has

---

[166] These are the foundations of the Western capitalist welfare state.

introduced the market of "ideological commodities" such as cars, refrigerators and televisions, intended to satisfy the needs of standardized consumers. In this generic and faceless society, various consumer associations, trade unions, and political organizations have risen that unite the generic, anonymous individuals to defend their rights against the strong corporations. The state provides generic individuals with standardized collective services such as education, health, transportation and various social services.

Mass production is not limited to physical commodities; it includes cultural "goods" such as books, theater, films, art in general, and various pastime products.[167] Systems of mass distribution make all that available to the mass consumers.

The interdependent system of mass production and mass consumption requires coordination between supply and consumer demand.[168] Lack of coordination may result in shortage or, on the contrary, in waste of resources.[169] For this, there are two alternatives: "the invisible hand" – market forces balancing supply and demand,[170] and "the visible hand" – Keynesian economics, in which the state is the master planner that coordinates between production and demand.[171]

---

[167] Jessop, "Fordism and Post-Fordism", 41-48.

[168] Ibid.

[169] Kevin Robins and Frank Webster, *Times of the Technoculture: From the Information Society to the Virtual Life* (London: Routledge, 1999), 97.

[170] "The invisible hand" is a metaphor coined by the Scottish economist and philosopher Amad Smith in his book *The Wealth of Nations*.

[171] John Maynard Keynes, *The General Theory of Employment, Interest and Money* (London: Palgrave Macmillan, 1936). This book, influenced by and written during The Great Depression, introduces the Keynesian economic theory (also called Keynesianism), which became the basis of the macroeconomic model. Keynes' basic argument is that the Great

Advertisement, in its modern capitalist form, plays an important role in coordinating supply and demand: it informs the consumer of the available products and their prices and helps the producer market the goods it produces (or plans to produce), predict demand and adjust the scope of production accordingly. In the "invisible hand" system, the role of advertisement as a synchronizing mechanism is much greater; however, it is also significant in the "visible hand" system, as consumption cannot be forced – at least not in the quantities required to balance the supply.

Fordism Requires Education and Training

The education system in the industrial society (Toffler's second wave)[172] had a very specific role.[173] For one, it served as a babysitter that helped free parents from factory labor, but besides that, it prepared and trained the next generations for work at the assembly line. The Fordist capitalists realized they needed to train the uneducated masses of peasants to become good factory hands and effective participants in the industrial system. That need gave birth to the public school system, which has remained largely unchanged to this day. Its primary purpose was to produce adults prepared and fitted for real life in the industrial age.[174] Education and professional training, which had been part of the private

---

Depression cannot be explained using the classic equilibrium theory.

[172] Toffler, *The Third Wave*, 45-46.

[173] Walter Horace Bruford, *The German Tradition of Self-Cultivation: Bildung from Humboldt to Thomas Mann* (London: Cambridge University Press, 1975), vii.

[174] For a more detailed description of the requirements from the education system, see the discussion of Bildung in the next section.

domestic sphere before the industrial revolution, now became part of the public sphere.

The industrial society, as defined above, has shaped the modern-day public sphere and looks at individuals from the outside. From the industrial viewpoint, the public sphere is where anonymous generic individuals act and are acted upon, work, learn and consume physical and spiritual commodities. Large companies regard the individual as a resource – not much different from raw materials, energy or money – which they can use to achieve their goals. Trade unions may represent and defend workers' interests, yet they do not see distinct individuals; instead, they see a mass of generic people united by a common interest (collective labor agreements are a good example of that). The same goes for consumer associations and political organizations, which consider groups of people rather than individuals, and for the state, which, for the sake of objectivity, considers an implied average citizen, not real individual people. In other words, none of the prominent entities in the public sphere knows the individual personally – or has to in order to do its job. The education and professional training system has moved out into the public sphere, with standardized curricula and students who, most of the time, are generic and faceless, too. Individuals in the public sphere are like generic atoms in a gas tank, running around and occasionally bumping into each other for their entire lives. However, one thing those generic individuals have is the right to choose what they want from the general market of commodities. Thus, the advent of mass production and Fordism has redefined the public sphere and consequently increased individuality, but at the same time, limited it to the sphere of the nuclear family and the ability to choose which mass-produced commodities to consume.

Democracy as an Individual-Centered Political System

A democratic regime is centered around the individual and their right to elect, to be elected and to share in the responsibility of running a state. The question of whether democracy is the obvious consequence of universal human qualities, i.e., whether every human being possesses the qualities and abilities that make them fit for participation in political life, has been discussed since the days of ancient Greece. In Plato's *Protagoras*, the sophist Protagoras defends democracy by recalling a myth in which Zeus distributes shame and justice to every human in equal amounts, thus saying every human being possesses the sense of justice and morality needed in order to make political decisions.[175] Protagoras' theory is both individualistic and universal. Individualistic because it focuses on the single individual, and universal because it suggests that the human virtues required for democracy are God's gift to every human being without exception. Both these characteristics have pertained to the liberal democratic state.

Lior Rabi introduces a different view in his book *Omes ha-individualiyut* [The Burden of Individuality].[176] He believes that the rise of democracies made it necessary to educate the masses of peasants who had joined the political system and were suddenly required to take part in momentous decisions of war and peace in order to give them political awareness of what was going on globally, not just in their immediate surroundings. This resonates with the philosophy of **Bildung** – a

---

[175] Alexander Yakobson, "Ma mevin nagar be-politica? Ha-democratia ve-hatzdakata be-mashal Protagoras" [What Does a Carpenter Know About Politics?: Democracy and its Justification in Plato's 'Protagoras'] (in Hebrew), *Zmanim: A Historical Quarterly*, 64 (Fall 1988), 23-33.
[176] Rabi, *Omes ha-individualiyut*, 31-32.

tradition of self-cultivation through education and culture, which developed in Germany in the 19th century.[177] The goal of Bildung is the harmonization of one's mind, heart, selfhood and identity, and it preassumes that one has the ability to reach that goal. Bildung is the process and the result at the same time and is associated with concepts such as freedom, emancipation, and causality. It requires both communication with the world around one and withdrawal from it in order to allow the development of selfhood. The classic theory of Bildung emphasizes the dialectic relationship between objective and subjective, between inside and outside, and the belief that neither one can exist without the other. Selfhood is shaped both by the material forces acting on one from without and by inner processes of interpretation, education and learning. Education in Bildung is more than just memorizing material and gaining knowledge or skills. It is a lifelong process of personal development fed by the literary and cultural legacy of previous generations.

Either way, democratic liberalism is based on the belief that every human being is capable and competent for being a useful and equal player in the political game – whether that capacity is naturally present from birth or must be taught at a young age. A democratic regime is centered around the individual. **Liberalism** is centered around human rights and liberties and is the foundation for a democratic state. The role of the state in liberalism is to be "the watchman," i.e., to interfere only where the individuals fail to settle matters between themselves. In particular, it retains a monopoly on violence through institutions like the armed forces and police. Liberalism acknowledges no authority external

---

[177] Werner Jank, "Didaktik, Bildung, Content: On the Writings of Frede V. Nielsen", *Philosophy of Music Education Review*, Vol.22 (2), Fall 2014, 113-131.

to humanity; therefore, the basic authority is the law, before which everybody is equal. In a democracy, the line between the public and private spheres is considered from the viewpoint of the individual, from the inside out. The strongest manifestation of the belief that every individual is a rational and autonomous being is free elections – a process in which a physical space (the voting booth) is taken out of the public sphere and temporarily turned into a private space protected by law against any external interference. In liberal democracies, religion is also excluded from the public sphere and considered one's private business.

Taking the authority to use violence and force from the hands of the individual and transferring it to the state – or "the people" – also means that protecting you – either as an individual or as part of a social group – has become the responsibility of the state. This creates an inherent conflict between personal privacy and public interest. In modern society, individuals are required to give up some of their privacy – for example, to report their financial state to the tax authorities or to let the police into their private physical or virtual space (e.g., house or email account) in case of a suspected crime or danger to public safety. They allow the state – under certain circumstances – to conscript them for mandatory military service or to monitor them despite the clear violation of their autonomy and other elements of their personal privacy. The balance between personal privacy and the common good is set by every society based on its characteristics, as described in previous chapters.

On top of the basic recognition of the right to privacy and the necessary balance between personal privacy and public interest, there are laws regarding privacy protection and social norms designed to protect personal privacy.

Personal Privacy in the Modern Era According to the Three-Dimensional Space Model

Values: Every human being is an autonomous individual with clearly defined rights and obligations, which include human rights such as the right to property, free speech, protection from violence and abuse, civil society and free market.

Norms and regulations: A democratic "Fordist" society with a system of courts, laws and regulations to defend the value of human rights[178] and a government whose job is to maintain the balance between the interests of the capitalists and the working class. Industrialists, banks and trade unions are protected by laws and regulations. Individuals have civil duties (e.g., military service and tax payment). Instrumental education aligned with industrial goals.

In relation to privacy, Human rights include both material rights (e.g., property) and mental rights (autonomous individuals with a right to anonymity in the public sphere). Norms and regulations that define the interactions between individuals and between individuals and society.

Knowledge: Rapidly advancing scientific, mathematical and engineering knowledge that defines specific sciences and uncovers the laws of nature. The accumulated knowledge is translated into engineering achievements that support the system of mass production and mass consumption, as well as the urban liberal democratic society.

Spiritual culture: Philosophy that defines every human being as a rational, autonomous individual who has rights. A culture based on

---

[178] As demonstrated, for example, by Warren and Brandeis' article, or by the Fourth Amendment to the US Constitution.

the Fordist model of mass production and mass consumption. Instrumental education is not based on people's individuality but is designed to train children and young people for life in a Fordist society. For the system, individuals are faceless and generic.

In relation to privacy, The public and private spheres are well-defined and hardly overlap at all. Every human being is an autonomous individual who has human rights and property rights protected by laws and regulations. In the public sphere, the individual is anonymous and generic. Individuality is realized mainly in the private sphere.

Social culture: Representative democracy is the political regime. The democratic system has three independent branches – the legislative branch, the executive branch and the judicial branch. Laws and judicial systems to manage conflicts between individuals and between individuals and society. Popular sovereignty is manifested through the three branches of government that defend the people's rights and define their duties. A culture of mass production and mass consumption, in which the individual is an anonymous and generic object involved in "give and take" interactions in the mass society. The majority of social interactions are between strangers.

Personal privacy is defined by laws and social norms. In the public sphere, individuals are anonymous and generic – whether as consumers or as production workers. In the political sphere, individuals are anonymous, too, and their political rights are realized through elected representatives. Social culture is fed by the mass culture. Personal privacy as a manifestation of individuality and autonomy is fulfilled mostly in the private sphere, which largely overlaps with the domestic sphere.

Technological culture: A combination of "pure" and "applied" sciences supports a Fordist industrial system – a system of mass production that uses organic fuels. Formation of large megalopolises. Advanced agriculture frees farmers from industry labor. Mass media as a pillar of representative liberal democracy. The same technological culture also supports the existence of a private physical and virtual sphere in which people can fulfill their personal privacy.

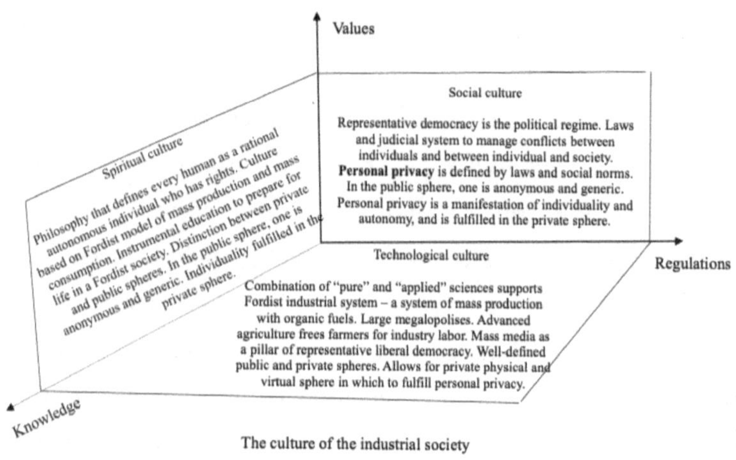

The culture of the industrial society

The main conclusion from this chapter is that modern culture has created two distinct, nearly untouching spheres: one is the private sphere, limited to one's inner world and intimate family, in which privacy has both a physical, material aspect and a mental and virtual aspect; the other is the public sphere, in which individuals are anonymous, but powerful when they unite as workers, consumers or participants in political life. Both spheres have legal and normative protection. In addition, it emphasized the inherent conflict between personal privacy and public interest (in particular, public safety).

Chapter 6

Personal Privacy – Identity and Private Space

The Components of Personal Privacy

Privacy is part of the private sphere, one's private space, and has many different aspects and components. The difficulty of defining it has been noted by many scholars, including Daniel Solove, who writes that "privacy is a sweeping concept, encompassing (among other things) freedom of thought, control over one's body, solitude in one's home, control over information about oneself, freedom from surveillance, protection of one's reputation, and protection from searches and interrogations."[179] In this chapter, I will try to clear the confusion and vagueness surrounding the term "personal privacy" by extracting the basic components of privacy from six conceptions which, according to Solove, capture the main ideas in the discourse about privacy. These conceptions are closely related and sometimes overlap, and together, they cover the idea of privacy in its modern Western liberal interpretation. The six conceptions are as follows: (1) the right to be let alone – Samuel Warren and Louis Brandeis's famous formulation for the right to privacy; (2) limited access to the self – the ability to shield

---

[179] Daniel J. Solove, "Conceptualizing Privacy", *California Law Review* Vol.90 (4) (July 2002), 1088-1154.

oneself from unwanted access by others; (3) secrecy – the concealment of certain matters from others; (4) control over personal information – the ability to exercise control over information about oneself; (5) personhood – the protection of one's personality, individuality, and dignity; and (6) intimacy – control over, or limited access to, one's intimate relationships or aspects of life.[180]

(1) **The right to be let alone:** This formulation first appeared in 1890 in the famous article by Samuel Warren and Louis Brandeis called "The Right to Privacy," which had an enormous impact on the US legal system.[181] At first glance, the article seems to be about the right to privacy rather than about privacy itself, but in fact, it is about the physical and mental aspects of privacy and the legal justification of protecting it: "Warren and Brandeis observed that increasingly, 'modern enterprise and invention have, through invasions upon his privacy, subjected [an individual] to mental pain and distress, far greater than could be inflicted by mere bodily injury'"; "According to Justice Douglas: [The] right of privacy was called by Mr. Justice Brandeis the right 'to be let alone.' That right includes the privilege of an individual to plan his own affairs, for […] every American is left to shape his own life as he thinks best, do what he pleases, go where he pleases."[182] Based on these interpretations, "the right to be let alone" can be summarized as one's right to be with oneself without external intrusions. The elements of privacy underlying this conception are the human body and the human mind, the basic units of selfhood, which have a right to be let alone. The right to be let alone applies to the physical, mental and virtual worlds alike; all three come

---

[180] Ibid.

[181] Warren and Brandeis, "The Right to Privacy".

[182] Solove, "Conceptualizing Privacy", 1100-1101. Quotation in source is from Warren and Brandeis, "The Right to Privacy".

under the definition of private space where one deserves to be let alone. Without a private space in this broad sense, there can be no privacy. However, the limits of the private space or spaces are not discussed in the article. Former President of the Israel Supreme Court Justice Aharon Barak called it "an autonomous unit" that envelops and follows one wherever one goes.[183] The basic components of privacy that can be extracted from this conception are body, mind, and private space (both physical and virtual).

(2) **Limited access to the self:** Ernest Van Den Haag (as quoted by Solove) describes it as the right to exclude others from watching, utilizing or invading one's private realm. However, this conception does not only include limiting the access of unwanted others to one's private space but also one's ability to control the extent of distance and seclusion and to decide who may enter the said private realm. E. L. Godkin talks about "the right of every man to keep his affairs to himself and to decide for himself to what extent they shall be the subject of public observation and discussion."[184] According to Prof. Ruth Gavison,[185] limited access to one consists of three independent and irreducible elements: secrecy (information about one), anonymity (attention to one), and solitude (physical access to one).[186]

One element of privacy underlying this conception is control over one's private space, including control over one's body and mind.

---

[183] High Court of Justice ruling 6650/04, Anonymous vs. Netanya Regional Rabbinical Court (published in November 2006).

[184] E. L. Godkin, "Libel and its Legal Remedy", *Journal of Social Science* 12 (December 1880), 69-80.

[185] Ruth Gavison, "Privacy and the Limits of Law", 433.

[186] As I will elaborate later on, I do not consider private information to be a separate component of privacy, but rather an essential and inherent part of all the other components.

Control over one's private space can be translated as individual autonomy, i.e., the ability to set for oneself the rules of entry into one's private space. Another element is anonymity – the right to remain unidentified. This conception does not clarify the boundaries of the private space or define the requirements for a private space "entry ticket," but it does clearly state that the private space has boundaries and that entry requires a "ticket" which only the individual in question may define as part of their autonomy. Thus, we can extract the element of control over the boundaries of and rules of entry into one's private space. However, this process is not unilateral: both the boundaries and the rules of entry are generally determined through social negotiation rather than by one's single-minded decision. This returns us to the argument of whether privacy has any meaning outside the social context.[187] To conclude, this conception contains the following basic components of privacy: body, mind, private space, autonomy, and anonymity.

(3) **Secrecy:** A view of privacy as secrecy and of the disclosure of secret information as a breach of privacy. The underlying concept here is control over one's secret information and over the time and form of its disclosure. Secret information may be related to aspects of one's life, one's body, deeds, thoughts, property, etc. The right to secrecy should be balanced against fair disclosure of personal information that the public has the right to know, also when regarded as a subset of limited access. A view of privacy as secrecy may suggest that information that is not secret cannot be private and that once it is disclosed, there is no more justification for privacy. Things done or seen in public cannot be considered private information because everybody has access to public

---

[187] Moore, *Privacy: Studies in Social and Cultural History*, 73.

spaces, and therefore, one cannot expect privacy there.[188] In Solove's words: "In a variety of legal contexts, the view of privacy as secrecy often leads to the conclusion that once a fact is divulged in public, no matter how limited or narrow the disclosure, it can no longer remain private."[189]

However, not all things private are also secret. The existence of sexual relations between husband and wife, for example, is no secret, yet such relations clearly belong to the private sphere. Thus, from the conception of privacy as secrecy, we can only extract one basic component of privacy – control over the transfer of information from the private sphere to the public sphere, i.e., autonomy over one's own body, mind and actions. In contrast to the limited access conception, which views the private sphere from the outside, here, the point of view is from the inside out.

(4) **Control over personal information:** According to this conception, privacy is the claim of individuals, groups, or institutions to determine for themselves when, how, and to what extent information about them is communicated to others. In Charles Fried's words, "Privacy is not simply an absence of information about us in the minds of others; rather, it is the control we have over information about

---

[188] I will discuss this issue further in the section on deep and general privacy. For now, I will only say that I disagree with this assumption, which does not make sufficient distinction between privacy and the right to privacy. If private information is publicly disclosed, this is a breach of privacy, but not a waiver of the right to privacy. I also disagree with the dichotomous distinction between secret and non-secret. I believe there are different levels of secrecy, from which derive different levels of privacy.

[189] Solove, "Conceptualizing Privacy", 1107.

ourselves."[190] Richard Parker offers a much broader definition of privacy: "control over when and by whom the various parts of us can be sensed by others" – i.e., seen, heard, smelled or touched.[191] However, personal information is a broad and complex concept that includes facts, such as one's history and behavior patterns, in addition to expressions of the self, such as feelings and thoughts, and the two are not always distinguishable. Moreover, personal information is often formed in relationships with others, in which case it is not controlled entirely by the individual in question but shared with the other party or parties to the relationship, and sometimes with other human or Internet entities (e.g., Google) as well. This conception of privacy also fails to address elements of personal information that are disclosed involuntarily, such as body odor or behavior patterns. According to Alan Westin, personal information should be defined as a property right.[192] This approach is based on the philosophy of John Locke, who stated that "every man has a property in his own person." Locke extends the idea of self-ownership to also include the products of one's labor, including intellectual products such as thoughts and ideas.[193] Therefore, in Westin's interpretation, control over personal information is regarded as a subset of intellectual property rights.[194] The conception of privacy as control over personal

---

[190] Charles Fried, "Privacy", *Yale Law Journal* Vol. 77 (1968), 482-483.

[191] Richard B. Parker, "A Definition of Privacy*", *Rutgers Law Review* Vol. 27, Issue 2 (1974), 281.

[192] Alan Westin, *Privacy and Freedom* (New York: Athenum, 1967), 324.

[193] John Locke, *Second Treatise of Government* (London: Awnsham Churchill, 1689), 27.

[194] The extension of property rights to information will be discussed later on in greater detail. Unlike other types of property, possession of information by one person does not make others unable to possess or consume the same information.

information can be viewed as a subset of conceptions (2) and (3), yet its interpretation does add the basic components of private property (personal information as property), information about relationships with other entities (the nature and intensity of one's social relationships), information as an extension of the self, and involuntarily disclosed information that is beyond one's control. This conception does not define which personal information should be considered private and which should not, though it does distinguish between the two. Nor does it discuss the need to balance the right to privacy against the public right to information.

(5) **Personhood:** According to Solove's description, this conception views privacy as a way of protecting personhood – "those attributes of an individual which are irreducible in his selfhood."[195] It can be divided into two categories: a) individuality, dignity, and autonomy. And b) anti-totalitarianism.

a. Individuality, dignity, and autonomy: One's individuality, human dignity, and autonomy over one's body, feelings, opinions and actions are the key elements of one's personhood and, therefore, also parts of one's privacy.[196] Solove quotes an excerpt from a US court ruling declaring "the right of every individual to the possession and control of his own person, free from all restraint or interference of others, unless by clear and unquestionable authority of law."[197] The very knowledge that one is being observed, or worse – under surveillance, affects one's personhood, restricts one's personal space, injures one's dignity and limits one's autonomy in every aspect (including body, decisions,

---

[195] Solove, "Conceptualizing Privacy", 1116.

[196] To these I would like to add memories, which, some may say, are one's true essence. See further discussion later on.

[197] Ibid., 1117.

feelings and actions).[198] The awareness of being observed by others is strongly connected to the right to be let alone.

However, discussions of privacy as protection of personhood often confuse it with other concepts such as autonomy and liberty. Another problem is the lack of a clear definition of "personhood." Moreover, "dignity" is a much broader concept that is not limited to privacy. Solove quotes Gavison's example that "having to beg or sell one's body in order to survive are serious affronts to dignity, but do not appear to involve loss of privacy." This argument that objectification has nothing to do with privacy undermines the perception of autonomy over one's own body as part of one's privacy, for if my body is – permanently or temporarily – an object, it cannot be considered part of my privacy, as privacy clearly does not apply to objects. It also undermines the assumption that one's body is one's private and untradeable property. So, it seems that the conception of personhood as an element of privacy focuses more on why privacy is essential than on what privacy is. The basic components of privacy that can be extracted from this part are autonomy, anonymity and identity.

b. Anti-totalitarianism: The limits of the self and conflicts between one's self and the selves of others. No self-definition can be completely unrelated to other individuals or the sovereign because it is an element of social life. Personhood is a priori limited by laws and social norms. This part focuses on balance: on the one hand, personal privacy is limited to expressing oneself within the social norms and without infringing on the personhood of others, but on the other hand, it is protected from intrusion and coercion by the state. The conception of

---

[198] The meanings of surveillance and the ways ICTs enable it will be discussed in the following chapters.

privacy as anti-totalitarianism does not introduce any new components of privacy, but it does attempt to define its limits and set balances. Solove calls privacy "both a positive and negative right": not just a freedom from the state, but a duty of the state to protect individuals and their property through legislation and law enforcement.[199] In return, individuals willingly give up some extent of their autonomy and confidentiality of personal information. The basic components of privacy that can be extracted from this conception are autonomy – in its broadest sense of autonomy over one's body, thoughts, emotions, decisions and personal relationships; identity; and relationships with other entities. These basic components are dependent on context and social norms.

(6) **Intimacy:** The conception of privacy as a form of intimacy focuses not on the individual themself but on the different relationships – particularly intimate relationships – they have with other people. We form relationships with differing degrees of intimacy and self-revelation, and privacy is perceived as the ability to maintain the desired levels of intimacy for each of our varied relationships.[200] According to Jeffrey Rosen, intimate relationships require space safe from the gaze of the crowd; without it, it would be impossible to build the mutual trust that is essential for intimacy.[201] Since, as mentioned, only in one's private space can one truly be oneself, private space is essential for having intimate relationships. According to Fried, intimacy is the sharing of information about one's actions, beliefs or emotions with people one trusts.[202] The

---

[199] Ibid.

[200] Robert S. Gerstein, "Intimacy and Privacy", *Ethics* Vol. **89** (1) (October 1978), 76-81.

[201] Jeffrey Rosen, *The Unwanted Gaze: The Destruction of Privacy in America*, **8**.

[202] Charles Fried, *An Anatomy of Values: Problems of Personal and Social Choice* (Cambridge, Harvard University press, 1970), 142.

level of intimacy varies depending on the intensity (the emotional depth) of the relationship and on the sensitivity of the disclosed information. These two elements are connected but not necessarily in direct proportion: for instance, you can share certain information with your therapist but withhold it from your spouse. From this conception, we can extract the following basic components of privacy: intimacy of personal relationships – which can be extended to intimacy of relationships with confidants (including also, for example, one's healthcare provider or bank); control over personal information; and control over one's private space, in other words – body, mind, actions, relationships with other entities, private space, and autonomy.

From the six conceptions of privacy discussed by Solove, I was able to extract the following **basic components of privacy**:

Private space: The physical and virtual area where one has privacy. Private space has a static physical aspect, which is one's place of residence, and a dynamic physical aspect, which can be envisioned as an envelope or bubble that surrounds and follows one wherever one goes.

Body: One's body, which includes one's physical characteristics such as appearance, height, weight, strength, health, senses and medical conditions, as well as the brain and nervous system, which are often perceived as carriers of mental properties.

Mind: The abstract part of one's person, which includes one's thoughts, feelings, desires, fears, self-perception, memories, and other mental and psychological elements. As we saw earlier, only in the modern era was the concept of personal privacy extended to include the mental and psychological realm.

Actions: The things one does and the way one behaves in the world, including things one says or writes, one's physical movements, and one's social and intimate relationships (who one has social or intimate relations with, the nature, frequency and intensity of the relationships, whether they include expression of feelings and/or exchange of information, etc.) – in other words, all of one's everyday behaviors and acts, including those related to other people. One's actions and behaviors are also a manifestation and realization of personal decisions and, as such, are an expression of individual autonomy – or lack thereof, if one is not acting of one's own free will.

Property: One is the sole owner of one's body and mind; one's personal physical space is one's private area where one can do whatever one likes; ownership of the abstract products of one's mind, such as thoughts and memories.

External entities: Other players in the system of balances between the protection and violation of one's privacy. Such entities can be human – other individuals or political, social or commercial organizations (e.g., commercial companies or governments) – or they can be social networks, such as Facebook or Google. Some of them are one's confidants who possess confidential private information about one, for example, healthcare providers who possess information about one's physical and mental health, lawyers who possess confidential legal information, banks who possess confidential financial information, and intimate partners who possess intimate information about one.

Relationships with external entities: The nature of one's relationships with other entities (whether the relationship is intimate, familial, economic, legal, work-related, subordinate, professional or social), including the power relations between the parties and the intensity of the relationships (the importance attributed to the relationship by each of the parties, the emotional strength of the relationship, the extent of mutuality, the amount of time invested in it, etc.).

Autonomy: An individual's capacity for self-determination or self-governance, the ability to decide for oneself, without any external interference, regarding one's own body and mind.[203] Philosophers make a distinction between personal autonomy, moral autonomy and political autonomy. Moral autonomy, usually traced back to Kant, is the capacity to deliberate and to give oneself the moral law rather than merely heeding the injunctions of others; personal autonomy is the capacity to decide for oneself and pursue a course of action in one's life, often regardless of any particular moral content; and political autonomy is the property of having one's decisions respected, honored, and heeded within a political context. According to the liberal democratic approach, all three are part of an individual's privacy. The individual makes moral decisions for themself and is responsible for their consequences – good or bad; that includes faith and religious practices, which in modern society are considered people's private business. The individual interacts with and affects reality as an independent, autonomous agent, and representative democracy is based on the autonomy of individuals to be political agents who elect their representatives in the process of free and secret elections to ensure the autonomy and authenticity of their choices.

Identity: The philosophical question of human identity has concerned humanity, at least since the days of Plato.[204] It has been phrased in many ways,[205] such as "Who am I?" or the Socratic "know

---

[203] "Autonomy", Internet Encyclopedia of Philosophy https://iep.utm.edu/autonomy/ (Accessed 28/3/2024)

[204] Floridi, "The Informational Nature of Personal Identity", 549-566.

[205] Ilya Levin, "Cyber-physical Systems as a Cultural Phenomenon", *International Journal of Design Sciences and Technology*, Vol. 22 (2016), 67-80.

thyself" or Paul Gauguin's "Where do we come from? What are we? Where are we going?".

Identity is a vague and complex concept. In this book, I make a distinction between "selfhood" and "identity," which I define as the outer representation of one's inner self. According to the liberalist approach, selfhood is a product of one's unique individuality,[206] for even though every individual is made up of the same universal building blocks[207] such as common human physiology, genetics[208] and cognitive abilities, the particular combination of those, together with every person's unique history, is what creates each individual's unique selfhood. According to Luciano Floridi, the philosophy of information assumes that selfhood precedes identity.[209] Therefore, a person's identity is also unique, as it represents that person's unique selfhood. This does not mean that identity is static and unchanging; it is only that it can always be associated with one unique individual. In Chapter 2, I mentioned two psychological approaches described by Ferdinand Schoeman: one approach says that every individual person has an inner nucleus, a core self; for privacy reasons, people usually reveal only some parts of that nucleus, while other parts are covered by different masks. The second approach says there is no "atomic" inner nucleus; the self has many faces, which are manifested in different circumstances.[210] This means an individual can have multiple identities, each of which either represents a different part (or parts) of the same "core self" or an altogether different self or selves.

---

[206] Rabi, *Omes ha-individualiyut*, 255.

[207] Garcia, *Individuality: An Essay on the Foundation of Metaphysics*, 234.

[208] Similar to the way every human's DNA is made up of the same four basic units, but each person's complete DNA is unique.

[209] Floridi, "The Informational Nature…", 2.

[210] Schoeman, "Privacy and intimate information", 410-411.

Raffaele Rodogno defines several types of personal identity, which may all represent the same selfhood.[211] The different identities are time- and context-dependent and not always distinct from each other. Some of them may exist simultaneously, while others replace previous identities. Rodogno lists the following types of identity, or "identity questions": (1) Passport identity: information that can be found in identification documents (e.g., ID card, driver's license, passport). Before the age of ICTs, the connection between a passport identity and selfhood was made by means of a photo; today, we also have means of biometric identification, as well as identification by IP (Internet protocol) address of a computer or mobile phone. (2) Numerical identity: identity based on one's cognitive abilities. It can change completely and suddenly, e.g., after a stroke. (3) Attribution identity: identity defined by various characteristics of the individual, such as occupation, marital status and place of residence. This type of identity is dynamic and changes in accordance with the circumstances of one's life. On the one hand, it is a manifestation of selfhood, as it reflects the individual's feelings, aspirations and choices; on the other hand, it constantly reshapes the selfhood because any discrepancy between one's selfhood and a particular attribution identity affects one's self-perception, decisions and actions. (4) Social function identity: the way one is perceived in the public sphere. It may be defined by one's occupation (e.g., opera singer), by one's social status (e.g., a rich white man), etc. This type of identity usually represents only a minor part of selfhood, but it plays a major role in shaping it and greatly affects one's interactions with other people. (5)

---

[211] Raffaele Rodogno, "Personal Identity Online", *Philosophy and Technology* Vol.25 (3) (2012), 309-328.

Attachment identity: Defined by what is important to the individual and what they care about.

Each of these identity questions represents some part of selfhood; their relative weight and significance change according to the circumstances, similar to the previously described theories of partial exposure of the self or exposure of one of the self's many faces. However, despite the multiplicity of identities and the fact that they may be very different from one another, they can all be associated with one physical entity with a continuous memory and experience.[212] That physical entity, which contains the selfhood, is unique, and therefore, according to both previously described theories of selfhood, all the identities are outer representations of the same selfhood.

All these are identities that exist in the real world.[213] Practices for maintaining multiple identities have evolved long before the age of ICTs: for instance, publication of written works under a pseudonym was quite common in the Anglophone world in centuries XVII-XXIX, inter alia, among American politicians who wished to keep their anonymity and privacy.[214] It became possible due to the transition from face-to-face interactions based on personal acquaintance to written interactions based on the printing press – journals, books and newspapers. Those interactions in the dawn of the modern era were interactions between

---

[212] Marya Schechtman, "Stories, Lives, and Basic Survival: A Refinement and Defense of the Narrative View", in: *Narrative and Understanding Persons*, ed. Daniel Hutto (Cambridge: Cambridge University Press, 2007), 155-178.

[213] As opposed to online identities, which exist only in the virtual world.

[214] Eran Shalev, "Ha-masekhot noflot: aliyata shel ha-democratia ha-americayit ve-ovdan ha-pratiyut ba-siyakh ha-politi" [The Rise of Democracy and the Decline of Republican Privacy in the United States], *Zmanim: A Historical Quarterly*, 104 (Fall 2008), 34-43 (in Hebrew).

large groups of people who were strangers to each other but shared common interests and views – somewhat similar to today's virtual communities on social media.[215]

Another practice that allowed a multiplicity of identities was masquerades.[216] At a masquerade, one can temporarily assume an identity very different from one's own, including gender swapping or assuming an inhuman identity. Adopting a different gender identity has become legitimate in modern society,[217] to the point of replacing one's birth sex completely, thanks to the developments of modern medicine. A famous literary example of a person with two distinct identities is Robert Louis Stevenson's novella *Strange Case of Dr. Jekyll and Mr. Hyde*.

Thus, even before the age of ICTs, the multiplicity of identities could be divided into two categories: linear multiplicity – i.e., when an old identity is replaced by a new one,[218] and simultaneous multiplicity – i.e., when the same person has more than one identity at the same time.[219] The regulatory way of preventing illegitimate multiplicity of identities was to issue identification documents that would associate a person with their unique identity by means of a profile picture or fingerprints.

The new category of identities that was born in the age of ICTs can be collectively called "online identities." Those are identities one creates or assumes in the virtual world. Every person who is connected

---

[215] Howard Rheingold, *The Virtual Community: Homesteading on the Electronic Frontier*, 2nd ed. (Cambridge, Mass: MIT Press, 2000), 1-18.
[216] Wahrman and Farkash, "Im ein ani li", 62-71.
[217] LGBTQs received social and legal recognition around 30 years ago.
[218] E.g., in witness protection programs the witness is given a new identity to protect them from the people or organizations they are testifying against.
[219] E.g., spies in an enemy country may assume a cover identity thar exists in parallel with their real one.

to the Internet has at least two identities at the same time: their offline identity and their online identity.[220] Online identities can be divided into two sub-categories: identities one creates for oneself on social media and in various Internet applications and user identities (profiles) created by large Internet companies to use as a business aid.

As far back as 1995, Sherry Turkle described multi-user domains (MUDs) as multiplayer virtual worlds in which the users themselves create the world in which they operate and design online identities (avatars) for themselves.[221] This may seem similar to assuming a fake identity at a masquerade, except that the virtual avatars in MUDs can "live" as long as the offline identity is not limited by time and space (unlike a masquerade) and can have all the identity types defined by Rodogno (passport identity, numerical identity, attribution identity, social function identity, and attachment identity). This makes those online identities more similar to real-world, offline identities. One's online identity may have similar characteristics to one's offline identity, or be completely different, or anything in between. The level of association between the online and offline identities can vary greatly.[222] While behind the mask at a masquerade, there is a very real person with distinct traits we can capture with our senses and thus identify them, an online identity may have nothing in common with any real person. Online identities are often used in social media applications like Facebook and dating applications like Tinder. People often create more than one online identity – mainly to remain anonymous but also to

---

[220] Floridi, "The Informational Nature of Personal Identity", 3-22.

[221] Sherry Turkle, *Life on the Screen: Identity in the Age of the Internet* (New York: Simon & Schuster, 1995), 111-119.

[222] There is an "arms race" of trying to associate online identities with real individuals, which will be discussed in the section on anonymity.

present different identities to different audiences or different partners. Online identities can certainly be false and are sometimes created with a specific purpose to deceive or to "contaminate" the public sphere: for example, publishing photoshopped images of models presents a false body image that young women may wrongly see as an ideal to aspire to.

Online identities created by Internet apps match the choices, experiences, and actions of specific real individuals and change as the apps collect more and more information about the individual in question. Many times, e.g., on Tinder, Amazon, Facebook and other similar applications, they are added to the online identity the individual knowingly created.

Both sub-categories of online identities have the following characteristics:[223] (1) Detachment between location and presence: the presence of the online identity is not limited to any physical location or physical entity; it is present independently in the virtual space, which has no space coordinates. (2) Detachment from memories and interactions: in the real world, our identity is shaped by our experiences, memories and interactions with the world. However, an online identity one creates for oneself may be completely detached from all those and be shaped without any experience at all, though online identities created by Internet apps are based on the experiences of real-life individuals. (3) Detachment of identity shaping from self-contemplation and contemplation of one's relationships with others (trying to see oneself through others' eyes), which is an essential part of offline identity shaping. (4) Online identities are shaped by contemplating a self that may be completely imaginary. The question "Who am I for you?"

---

[223] Floridi, "The Informational Nature of Personal Identity", 18-20.

becomes "Who am I online?"[224] The online identity becomes a reflection of one's identity as it is built and aggregated by Internet apps.

Besides online identities that belong to real individuals, there are chatbots – software programs designed to chat with humans online and mimic human interaction as closely as possible. Despite being artificial entities, these online identities have a "life" of their own in cyberspace.

However, despite the ability to have multiple online identities at the same time and to make them very different from one another, they can all still be associated with and traced back to one physical person with a continuous memory and experience.[225] In the age of ICTs, it is done by means of constant monitoring of people through IoT technologies. Neither of those systems alone is adequate for making the associations required in order to trace all the different identities back to a specific individual person, but using several systems to cross-reference data can reveal the specific individual behind the multiple online identities.[226] The information collected by ICTs is aggregated in interconnected databases and analyzed using advanced technology.[227] As we will see, this is an actual violation of personal privacy: biometric data pools add up physical and mental properties and characteristics, enabling the association of various online profiles with specific people. The combined use of technological means of surveillance and monitoring eventually renders a multiplicity of identity practices useless.

---

[224] Ibid.

[225] Schechtman, "Stories, Lives, and Basic Survival...", 162-167.

[226] For a detailed explanation, see the section on k-anonymity in Appendix C.

[227] The specific practices will be explained later.

Anonymity: As Michael Birnhack has stated, anonymity is a concept that can be hard to define. The literal meaning of the word "anonymous" is "nameless," and it usually means doing something without identifying oneself.[228] In his book, Birnhack reviews and discusses the various definitions and categories of anonymity.

Helen Nissenbaum describes anonymity as the inability to get to (or at) a person, i.e., to associate elements of information or things done in the physical world with a specific individual.[229] She thus extends the concept of anonymity, which before the age of ICTs meant conducting oneself without revealing one's name (e.g., a literary work whose author is unidentified, an anonymous donation, or conducting oneself in a foreign city where nobody knows one's name), i.e., preventing access to one's passport identity.[230]

According to Andreas Pfitzmann and Marit Hansen, the anonymity of a subject means that the subject is not identifiable within a set of subjects (the anonymity set).[231] This is a further extension of Nissenbaum's definition because it also includes subjects that are not individual humans. This form of anonymity is weaker because it only means that a subject is unidentifiable within a given set, not unidentifiable in general. There is no general protection of the subject's identity, only the inability to associate a specific identity with a specific subject. This definition overlaps with Nissenbaum's in the sense of

---

[228] Birnhack, *Merkhav Prati: Ha-zkhut la-pratiyut ha-ishit…*, 269.

[229] Helen Nissenbaum, "The Meaning of Anonymity in an Information Age", *The Information Society,* Vol. 15 (2) (May 1999), 141-144.

[230] Rodogno, "Personal Identity Online", 310.

[231] Marit Hansen and Andreas Pfitzmann, "A terminology for talking about privacy by data minimization: Anonymity, Unlinkability, Undetectability, Unobservability, Pseudonymity, and Identity Management", 9. https://www.researchgate.net (Accessed 24/1/2021)

preventing physical access to the subject. The immediate conclusion from Pfitzmann and Hansen's definition is that anonymity is possible only if the subject is part of a group or a category; the bigger the set of subjects one is part of, the greater one's anonymity.

In the age of ICTs, the claim to anonymity has been extended to:

- Sending electronic messages (e.g., emails) without the sender's (and sometimes also the addressee's) name being revealed;
- Participation in virtual interactions (chat rooms, online gaming, online dating services, e-commerce) without one's real name being known to other participants;
- Purchasing goods and services online using virtual currency (e.g., bitcoin) and without giving one's name;
- Ability to visit any website without having to divulge one's identity.

Anonymity also means the ability to conduct oneself under a false identity (online or offline), which cannot be associated with a particular real-life individual. Before the age of ICTs, you could remain anonymous by preventing access to your passport identity; now, it is much more difficult: a few seemingly unrelated pieces of information that are aggregated in data pools shared and analyzed by AI software can suffice to reveal the identity of any individual. Those pieces of information may include your shopping habits, web surfing patterns, or the places you visit. In other words, through various fragments of indirect information, it is quite possible to physically reach a person – to knock on their door and demand money or sue them for something they have done or said. According to Bruce Schneier, in the age of ICTs, it is impossible to remain anonymous against the surveillance and monitoring conducted by governments, commercial companies and other entities because, in the physical world, one's actions are recorded by IoT sensors, while in the virtual world, all the patterns of one's online behavior – including web surfing patterns, tweets and interactions – are constantly monitored; thus, ICTs are making anonymity impossible.[232]

Anonymity is never a game of one: there is the individual who wishes to remain anonymous and the other who wishes – for whatever reason – to breach that anonymity. Most often, it is a game of three: the individual, their confidant(s), and the other. Before the age of ICTs, anonymity was a social norm protected by laws that were part of the Fordist culture. The age of ICTs has extended the concept of anonymity and made it harder to achieve. It has challenged the social norms that supported anonymity, especially anonymity in the public sphere. Anonymity as part of personal privacy is an individual's shield against incapacitating surveillance by the government or by other individuals. It allows free self-expression and independence of thought, feeling and action. In commerce, it protects one from those who might try to manipulate one or take advantage of one's weaknesses and whims.

Solove lists control over personal information as one of the six main conceptions of privacy (conception 4). I believe that personal information is not a distinct basic component of personal privacy, but rather, that each basic component of privacy carries elements of information. Together, the elements of information carried by all the basic components of privacy make up all the personal information about us. Control over personal information is certainly part of personal

---

[232] Schneier, *Data and Goliath*, 4-42.

privacy, but it is a part extracted from the basic components of privacy, which carry the said personal information.

Popper's Three Worlds Theory as an Extension of the Conception of Private and Public Spheres

The idea of a private sphere is a constant element in every conception of personal privacy and applies to the physical and virtual realms alike. This is the sphere where one may be alone with oneself or with intimately close people. Neither of the six conceptions discussed by Solove defines the boundaries of the private sphere because some of them focus on one's inner (physical and mental) world, while others focus on the without. Defining the boundaries of the private sphere is difficult for several reasons: (1) This is the sphere one only reveals to oneself and to the closest people – but sometimes not even that, because the unconscious is part of the private realm as well; (2) The multiplicity of definitions for personal privacy and for the right to personal privacy makes it difficult to delimit the private sphere; (3) The boundaries of the private sphere depend on social norms and legal definitions; (4) The ICTs' increasing ability to uncover the private sphere: the boundaries of the private sphere change not only in accordance with the legal definitions and social norms but also in accordance with the developments in the field of ICTs.[233]

In the antique world, the private sphere was identical to the physical private space owned or controlled by one. This idea, manifested in the proverb "my home is my castle," has pertained in the liberal

---

[233] Those changes are interdependent; in particular, the advancement of ICTs drives changes in norms in regulations. Birnhack, *Merkhav Prati: Ha-zkhut la-pratiyut ha-ishit…*, 57-88.

democratic Western culture and received legal protection both in the US Constitution and in Israeli law. However, it is a limited conception of the private sphere, as it restricts the private sphere to a static physical place. In the liberal democratic Western culture, the idea was extended from something static to something dynamic – in the words of Justice Aharon Barak, like an envelope that follows one wherever one goes, inseparable from the individual themself, a sort of private space within the public space.[234] Some examples include voting booths in secret democratic elections and fitting rooms in clothes stores. In the conceptions described by Solove, the private sphere also includes the virtual space, which contains one's feelings. All this makes drawing the boundaries of the private sphere extremely difficult.

In order to overcome some of the difficulties of mapping the private and public spheres, I will use Karl Popper's theory of Three Worlds, which he defines in his famous lecture:[235]

World 1 is **the physical world**: the world outside us, which we can perceive by our senses or by means of technological aids such as microscopes and telescopes (for things that are too small or too far to be perceived by the human eye). Some parts of this world are considered part of the private sphere, e.g., one's body and one's home. Knowledge about this world is objective.

World 2 is **the mental or psychological world**: It contains our feelings, thoughts, emotions, and all the other mental states and processes that are supposed to be known only to ourselves unless we choose to express them or share them with another person. This world is private and known as it truly is only to the individual themself, while

---

[234] Ibid., 92.

[235] Popper, "Three Worlds – The Tanner Lecture on Human Values".

some parts of it – like the unconscious – are not entirely known even to them. Until the age of ICTs, every human society, including those that are not reviewed here, believed this world to be entirely and unequivocally part of the private sphere. Knowledge about this world is subjective because it is based on the reports of the individual in question regarding things that belong to their inner world.

World 3 is **in-between**: The objects contained in it are objective because there is a general consensus regarding their meaning and significance. Most of World 3 lies beyond the private sphere. Even though some elements in it are considered part of the private sphere, e.g., patents and scientific discoveries that have not been made public for whatever reason, this is usually temporary: patent protection expires, and secret knowledge about Worlds 1 and 2 eventually becomes public. The primary characteristics of the Three Worlds are summarized in the following table:[236]

	Contents	Status	Examples	Examples related to personal privacy
World 1	Physical objects	Objective	Table, chair, steel	Body, private property
World 2	Mental experiences	Subjective	Pain, joy	Character traits, opinions
World 3	Knowledge	Objective	Mathematics, Biology	Privacy information in data pools

---

[236] Ben-Israel and Tabansky, "Mabat beintkhumi al etgarey ha-bitakhon be-idan ha-meyda", 21.

Based on this, I would like to define the private sphere as containing all of World 2 as well as parts of Worlds 1 and 3 that are considered to be private and personal. Furthermore, I will define deep personal privacy as World 2 and general personal privacy as those parts of Worlds 1 and 3 that are considered private. Together, all the combined information of both types of personal privacy – deep and general – constitutes personal privacy information.

The public sphere begins where the private sphere ends. Supposedly, everything we do in the public sphere is open and exposed to all. The public sphere is present only in Worlds 1 and 3. Everything in it can be perceived by our senses, either directly or with the aid of technological means.

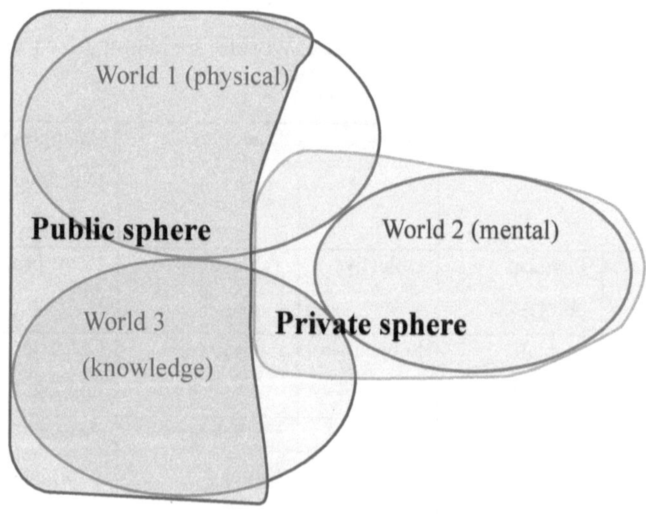

Relationship between Popper's Three Worlds and the private and public spheres

This definition of the private sphere demonstrates that it has static physical aspects planted in World 1 (e.g., one's home), as well as dynamic physical aspects that follow one wherever one goes.[237] Some of those dynamic aspects are physical and belong in World 1, e.g., one's body appearance, while others belong in World 2, e.g., feelings and thoughts.

---

[237] As per Justice Barak's definition.

Chapter 7

The Paradigm of Personal Privacy

What is a Paradigm?

Before I argue that the various conceptions of personal privacy can be framed together under a general "paradigm," let us, first of all, understand what "paradigm" is. According to Wikipedia, a paradigm in science and philosophy is "a distinct set of concepts or thought patterns, including theories, research methods, postulates, and standards for what constitute legitimate contributions to a field."[238] The Cambridge Dictionary defines "paradigm" as "a set of theories that explain the way a particular subject is understood at a particular time."[239] The term itself is attributed to science philosopher Thomas Kuhn. In his influential book *The Structure of Scientific Revolutions* (Kuhn, 1962), Kuhn defines paradigms as "universally recognized scientific achievements that for a time provide model problems and solutions to a community of practitioners." Aware of the unclarity of this definition, he explains in the postscript added in the second edition that he had been using the word in two different senses: one (he calls it the sociological sense) is all the beliefs, values, techniques, and so on shared by the members of a given community; the other is past achievements and solutions used as models or examples for solving the remaining puzzles.[240]

---

[238] https://en.wikipedia.org/wiki/Paradigm (Accessed April 03, 2024)

[239] https://dictionary.cambridge.org/dictionary/english/paradigm (Accessed April 03, 2024)

[240] Thomas Kuhn, *The Structure of Scientific Revolutions*, 2nd ed. (Chicago: University of Chicago Press, 1970), viii, 175.

A paradigm is a "matrix" composed of ordered elements of various sorts, each requiring further specification. All or most of the objects of group commitment that are part of the same paradigm are constituents of the disciplinary matrix, and as such, they form a whole and function together. Kuhn lists the following as the main components of a paradigm: (1) symbolic generalizations – the formal or the readily formalizable components that can be expressed in scientific, logical form, e.g., mathematical formulae or chemical reaction laws; (2) the metaphysical parts of paradigms – shared commitments to common beliefs, e.g., scientific consensus regarding physical objects and their properties; (3) values that define an adequate theory, adequate predictions and adequate explanations – the criteria to be used in judging whole theories and determining the accuracy level of proposed explanations and proofs (e.g., quantitative predictions are preferable to qualitative ones); (4) shared exemplars – concrete problem-solutions that serve as models of the way to solve similar problems.

The Paradigm of Personal Privacy and Its Components

The main conceptions of personal privacy formed by Western liberal democratic societies are all part of the same paradigm, which I shall call **the paradigm of personal privacy**. I will now break it down into the four main components proposed by Kuhn.

Symbolic Generalizations

The paradigm of personal privacy was mostly formed before or during the advent of ICTs; nevertheless, the following symbolic generalizations are at the base of the question of whether this paradigm is still relevant in the age of ICTs.

1. Moore's law.

2. The Internet.

3. The entropy function: $H(x) = -\Sigma\, p(x) \cdot log\, p(x)$. The entropy function measures the association between the degree of order or disorder in information and its randomness or uncertainty. It is a major tool used for data mining and for the prediction of facts based on data from big data pools.[241]

4. Big data: To illustrate the impact of big data pools on personal privacy, I will briefly discuss methods of data mining with decision trees,[242] which are based on the entropy function.

5. Cloud storage: The physical location of big data pools.

6. Cookies.

7. Profiling.

8. The Internet of Things.

9. *K*-anonymity: A formal model designed to protect the anonymity of users whose data are stored in a data pool (or multiple data pools). The basic concept of *k*-anonymity is that any given individual whose information is contained in a data release should be indistinguishable from at least *k*-

---

[241] Claude E. Shannon, "A Mathematical Theory of Communication", reprinted with corrections from *The Bell System Technical Journal*, Vol. 27 (July, October, 1948): 379-423, 623-656.

[242] Lior Rokach and Oded Maimon, *Data Mining with Decision Trees: Theory and Applications*, 2nd ed. (Singapore: World Scientific, 2014), 167-182.

1 other individuals whose data also appear in the release.[243] For example, *k=2* means that a given individual's data cannot be distinguished from the data of at least one other person.

10. Deep learning: A discipline in computer science that seeks to simulate the decision-making power of the human brain and plays a major part in AI technologies. What makes deep learning systems so unique is their ability to learn and improve themselves all the time, much like the human brain – the more the system is used, the more it "knows" and the better it works. Later on, I will discuss how deep learning technology can be used to violate privacy.

11. Data encryption: A range of mathematical tools and methods designed to keep information encrypted for anyone who is not authorized to know it. In the paradigm of personal privacy, these are the tools and methods of maintaining individuals' privacy.

12. Social distance: Some social distance seems to be necessary in order to maintain proper social relations.[244] I have categorized social distance as a symbolic generalization because I am proposing a metric for its calculation, in which I replace the term "social distance" with "extent of non-knowledge of personal information."

---

[243] Latanya Sweeney, "*k*-Anonymity: A Model for Protecting Privacy", *International Journal on Uncertainty, Fuzziness and Knowledge-based Systems*, 10 (5) (October 2002), 557-570.

[244] See the mathematical section in Chapter 2 for the proposed metric for calculating the "minimal radius" of social distance. This "minimal radius" is culture-dependent.

Without a tool for measuring social distance, it could be considered both a symbolic generalization and a metaphysical part of the paradigm.

Metaphysical Parts

1. Every individual human has a unique selfhood. The philosophical concept of selfhood is a product of individualism, which sees every human being as an individual in their own right who has the autonomy to fulfill their individuality, and which has given birth to the liberal democratic approach.

2. The existence of personal privacy, which (in its modern sense) is the product of the liberalist principles of liberty, equality and justice for all. Liberty defines the need for a private sphere where one can fulfill one's individuality; equality is the acknowledgment that all people possess rational ability and a set of universal rules of ethics that define the limits of individual autonomy and the social norms of distinguishing between private and public; justice defends the natural and universal human right of self-fulfillment. In the words of Bruce Schneier: "Privacy doesn't just depend on agency; being able to achieve privacy is an expression of agency."[245]

3. Our identity is a derivative of our unique individuality and of our right to fulfill it.[246] Even though every individual is

---

[245] Schneier, *Data and Goliath*, 126.
[246] Rabi, *Omes ha-individualiyut*, 255.

made up of the same universal building blocks, such as common human physiology, genetics and cognitive abilities, the particular combination of those, together with every person's unique history, is what creates each individual's unique selfhood. According to the liberal democratic approach, that unique identity must be protected. Identity has physical aspects, such as one's body, as well as virtual aspects, such as online identity.

4. Society: According to the Oxford English Dictionary, society is "people in general, living together in communities" or "a particular community of people who share the same customs, laws, etc.". In other words, society is the human environment in which the individual is an agent. It is the envelope that surrounds the individual, its boundaries and power determined through conflict and struggle with the individual, but it also complements the individual.

5. Private and public spheres: The private sphere consists of the physical and virtual areas where one has the right to fulfill one's self-potential, while in the public sphere, one is bound by the rules and norms of society. The line that separates the private and public spheres is culture-dependent and vague.

6. The individual as an agent: An individual human can be regarded as an agent that interacts with and affects the world, meeting all three requirements of agency: having an individual distinguishable identity, being a source of action, and acting according to norms.

7. Social distance: According to Georg Simmel, social distance is the extent of familiarity or unfamiliarity between people,[247] which constitutes the sphere of personal privacy. Some social distance seems to be necessary in order to maintain proper social relations.

8. Personal information: **Personal information** is information that can be univalently attributed to and associated with a specific individual, while **private personal information** is information that can be univalently associated with a specific individual and is inaccessible to others.[248] Both the American and European legal systems were aware of this difference between **personal information** and **private personal information**,[249] but while American law only regards information defined as private in this book as private information, European law regards any information defined here as personal as private information. This difference in approach is of great significance for the formation of a new paradigm of privacy.

9. Information protection laws help regulate the use of personal information while maintaining its privacy. These laws are based on the values of personal privacy formed by the Western liberal democratic civilization and are a manifestation of the right to be let alone as formulated by

---

[247] Simmel and Wolff, *The Sociology of Georg Simmel*, 312.

[248] In the case of private personal information, inaccessible also means "impermissible to others".

[249] Birnhack, *Merkhav Prati: Ha-zkhut la-pratiyut ha-ishit…*, 78.

Warren and Brandeis;[250] the conception of privacy as the right to control over one's privacy, including one's private information – as defined by Alan Westin;[251] Gavison's conception of privacy as limited access,[252] which consists of secrecy, anonymity and solitude; and Nissenbaum's conception of privacy in context.[253]

Based on all the above, in 1980, the OECD published eight principles[254] that were intended to provide the normative basis (expected norms of behavior) for information privacy protection laws and regulations:

- Collection limitation principle: Personal data should be obtained by lawful and fair means and, where appropriate, with the knowledge and consent of the data subject.
- Data quality principle: The collected data should be relevant to the purposes for which they are to be used and should be accurate, complete and kept up-to-date.
- Purpose specification principle: The purposes for which personal data are collected should be specified in advance.

---

[250] Samuel D. Warren and Louis Brandeis, "The Right to Privacy", in: *Philosophical Dimensions of Privacy: An Anthology*, ed. Ferdinand D. Schoeman (Cambridge: Cambridge University Press, 1984), 75-103.
[251] Westin, "The origins of modern claims to privacy", 56-59.
[252] Gavison, "Privacy and the Limits of Law".
[253] Nissenbaum, "The Meaning of Anonymity…", 141-144.
[254] OECD Guidelines on the Protection of Privacy and Transborder Flows of Personal Data, 1980.

- Use limitation principle: Personal data should not be disclosed, made available, or otherwise used for purposes other than specified, except with the consent of the data subject or by the authority of law.
- Security safeguards principle: Personal data should be protected by reasonable security safeguards against unauthorized access, loss, destruction, modification or disclosure.
- Openness principle: There should be openness about the identity of the data controller and its practices and policies with respect to personal data.
- Individual participation principle: The individual who is the data subject should have the right to access their data and to have it amended, rectified or erased (the right to be forgotten).
- Accountability principle: A data controller should be accountable for complying with these principles.

These principles have become the basis of personal information privacy-related laws and norms of behavior accepted in Western liberal societies. Paradigmatically, they can be seen as expressions of the basic principles of **consent** and **control**. The guiding legal principle is that you should give express consent to the collection of information about you. For this reason, Internet companies are required to obtain customers' advance consent (express or implied) to the collection of their information – by means of cookies or otherwise. Such consent should be conscious, i.e., you should be aware that you are giving your consent.[255]

---

[255] Regulation (EU) 2016/679 of the European Parliament (the GDPR), Recital 32, gives a broader definition of conscious consent and sets the

The requirement for consent also includes the second principle of control over information because by consenting or not consenting, you also choose which categories of information about you may or may not be collected.

According to most applicable laws, the consent for information collection is not sweeping but limited to those categories of personal information that the data subject has agreed to share with the data controller (the entity that collects the information) and to the strict purpose for which the information is being collected. In other words, the use of the information is limited, and there is some element of control over the information. The information must be obtained by fair and reasonable means, with the knowledge and consent of the data subject. The basic principles of consent and control include the element of solitude, which is a manifestation of the right to be left alone; the right to access one's personal information stored in any database and to ask for its rectification or erasure; the element of secrecy and confidentiality; and the accountability of the data controller for complying with these principles and maintaining these rights.

Values

1. Universal right to self-fulfillment: According to the liberal democratic approach, every person is a unique individual

---

following conditions: "Consent should be given by a clear affirmative act establishing a freely given, specific, informed and unambiguous indication of the data subject's agreement to the processing of personal data relating to him or her, such as by a written statement, [...] choosing technical settings for information society services or another statement or conduct which clearly indicates in this context the data subject's acceptance of the proposed processing of his or her personal data".

who has the right and duty to realize their full potential. This conception of every human being as an individual with liberties and rights is reflected in John Stuart Mill's words: "The individual is not accountable to society for his actions, in so far as these concern the interests of no person but himself."[256]

2. In liberal democratic societies, the behavior of individuals and the society as a whole is guided by the liberalistic principles of liberty, equality and justice for all.

3. Personal privacy in its modern Western sense is the way to mitigate and resolve the seemingly unsolvable tension between the private and the public by distinguishing the private sphere, in which one is free to realize one's full potential, from the public sphere, in which one is bound by the rules and norms of society.

4. The duties of individual and society: The individual must realize their individuality, and society must allow them to do that.

5. Balance between individual and society: The a priori assumption of the liberalism that arose from the American and French revolutions is that the inherent conflict between individual and society is unsolvable, and therefore, there is a need to find the appropriate social and political balances within which people can fulfill their individuality. Self-fulfillment is not unlimited. The proper balance must be found between one's right to privacy and other rights, such

---

[256] Mill, *On Liberty*, 178.

as the right to life and health.[257] Therefore, the line between individuality and universality is blurred.

6. Every adult individual is capable of and competent for being a useful and equal player in the political game – whether that capacity is naturally present from birth or must be taught at a young age.

7. The state as "the watchman": The state only interferes where the individuals fail to settle matters between themselves. In particular, it retains a monopoly on handling violence through institutions like the armed forces and police. Liberalism acknowledges no authority external to humanity; the authority is the law, before which everybody is equal.

8. Anonymity in the public sphere: In the public sphere, one is anonymous and generic. Political and economic entities do not need to know the specific individual; what they consider an implied average citizen is non-existent and, therefore, anonymous. Part of personal privacy is the right to protection of one's anonymity in the public sphere.

Exemplars

The following are used as examples and models of issues related to personal privacy:

---

[257] This conflict became evident during the COVID-19 pandemic, when severe limitations were imposed (though temporarily) on rights related to personal privacy, e.g., requirements to provide health information, mandatory testing, and so on.

1. According to Genesis 3 (6-11), Adam and Eve rushed to cover their nudity as soon as they became sentient after tasting from the Tree of Knowledge. Circa 1000 CE, Jewish Halakhist Rabbeinu Gershom Me'Or Hagolah instituted a ban on reading other people's mail without permission.

2. The birth of privacy as a legal right, in the modern sense of the right to be let alone, is attributed to Warren and Brandeis's revolutionary article.[258] The article examines the right to privacy in light of technological innovations such as photography and mass media (newspapers), which violated anonymity and privacy, making private information public or intruding into private space. The issues discussed in it are the right to intellectual property and the right to a private space free from external intrusions and safe from surveillance and control over one's actions, thoughts and feelings by others.

3. The publication of classified NSA documents by Edward Snowden in 2013[259] proved that the NSA had been spying on US citizens and political leaders around the world by electronic means using information collected and stored by Internet companies such as Google, Facebook and Apple. The documents leaked out by Snowden exposed violations of personal privacy by individuals all over the world, as

---

[258] Warren and Brandeis, "The Right to Privacy".

[259] According to The Guardian, the US government had been spying on embassies and representatives of 38 countries on US soil. https://www.theguardian.com/world/2013/jun/30/nsa-leaks-us-bugging-european-allies (Accessed 5/4/2024)

well as collaborations between governments and giant Internet corporations in violation of personal privacy.

4. The leaking of the personal data of 50 million Facebook users by Cambridge Analytica Ltd. to be unfairly used in Donald Trump's 2016 presidential campaign.[260]

5. The EU General Data Protection Regulation (GDPR) – a law that was designed to protect private information and includes severe sanctions for violators. The GDPR is a prominent example of privacy protection legislation passed by many countries, including the Israeli Protection of Privacy Law, which became effective in May 1981.

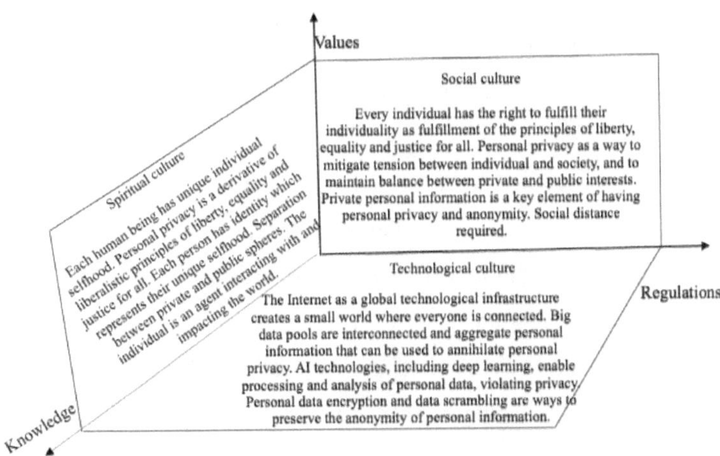

The privacy paradigm according to the three-dimensional space model

[260] https://www.nytimes.com/2018/04/04/us/politics/cambridge-analytica-scandal-fallout.html (Accessed 5/4/2024)

Chapter 8

Personal Privacy Information

Personal Privacy Information

Personal information is not a distinct component of privacy; rather, each of the basic components of privacy carries elements of information.[261] All the components of personal information combined together make up all of one's personal information. In this chapter, I will discuss the different types of information that are stored in data pools nowadays and the differences between information storage methods before and during the age of ICTs.

Before the age of ICTs, information was stored in analog formats – written (whether handwritten or typewritten), photographic or recorded – and could be perceived and understood by direct use of our senses, mainly sight and hearing. There was no shared platform, which made it difficult to connect different pieces of information together. In the age of ICTs, analog storage was replaced by digital storage. The digitization of information and its storage in large data pools, which are all connected to each other through a shared platform – the Internet – allowed us to make connections between different pieces and categories of information without dependence on their physical location and often without dependence on time. This change had an enormous impact on

---

[261] The basic components of privacy were discussed in detail in Chapter 6. This chapter will focus on the elements of information carried by each of those components.

the distribution of information in general and the distribution of personal information in particular. Nowadays, personal information about every human being around the world is being stored and aggregated. Information can be **viewed** in its digital form and **analyzed by machines without any need for human senses**. It is distributed in multiple copies and can be analyzed and understood without any help from human eyes or ears.[262] Moreover, up until recently, most personal information had a limited "life span" that roughly corresponded with the lifetime of the data subject, but now, information can survive much longer than the individual it refers to.

We have defined the **private sphere** as the physical and virtual areas in which one can have privacy. Now, let us redefine it in terms of information and knowledge: **deep personal privacy** is the information about one (one's body, mind, etc.) that nobody except one knows, i.e., the knowledge one has about oneself minus the knowledge the world (or society) has about one. **General personal privacy** is the information about one that is shared by one and one's confidants but not by the world in general (other individuals, databases, enterprises or organizations) – in other words, the knowledge one and one's confidants have about one minus the knowledge the world has about one. Confidants have an understandable commitment not to disclose this knowledge publicly and to keep it in the private sphere – or they would not be confidants. A confidant may be an individual – e.g., a doctor, a psychotherapist or a family member – or an institution, e.g., an insurance company or a bank.

---

[262] The transition from information collection by people (human net) to information collection by machines in government and military organizations is a proof of that.

Those are the formal definitions of the private sphere in terms of information and knowledge, or what Westin calls "social distance."[263]

Deep personal privacy is contained entirely in Popper's World 2, while general personal privacy includes parts of Worlds 1 and 3. Information about World 2 – the mental world – is not available by direct observation of that world but is inferred based on information from Worlds 1 and 3: conscious reports by means of speech, writing or art (drawings, music, dance); unconscious reports such as physical symptoms (odor, sweating, movements or body temperature); or indirect reports (e.g., family pictures which inadvertently reveal information about the relationships in one's family). That information is analyzed based on science and knowledge, which are part of World 3.

Gary Marx makes a somewhat similar distinction between different types of individual information, which he arranges as "concentric circles of individual, private, and intimate and sensitive information" with "information that is individual, private, and intimate and sensitive with a unique core identity at the center."[264] The outermost circle is "individual information" – "any data/category which can be attached to a person"; the next circle refers to private information, which includes "information about the person not known by others whose communication the individual can control as existentially private"; the next one is that of intimate and/or sensitive information – which is information that is not routinely known, is selectively revealed or is legally protected; the two final circles at the center, unique and core

---

[263] See Chapter 2 for a proposed metric for calculating it.

[264] Gary T. Marx, "Varieties of Personal Information as Influences on Attitudes Toward Surveillance", in: *The New Politics of Surveillance and Visibility*, eds. Kevin Haggerty and Richard Ericson (Toronto: University of Toronto Press, 2006), 79-110.

identification, concern a person's self-definition and the ways a person is perceived by others.[265] Marx gives multiple examples of information elements that belong to each of the circles; however, I cannot accept such dichotomous division. Some elements of information that he defines, for example, as intimate – such as the appearance of private body parts – may be exposed under certain circumstances allowed by the specific culture's norms and, therefore, belong partly in the sphere of deep personal privacy and partly in the sphere of general personal privacy. The same is true for each of the other circles. No information element can be placed exclusively in the sphere of deep personal privacy, general personal privacy or the public sphere.

The relationship between deep privacy, general privacy and personal information in general in the age of ICTs can be represented by the following graph. The outermost oval represents all the personal information about an individual, the dashed oval represents the personal information known to confidants, and the dotted oval represents the personal information that is known to the general public.

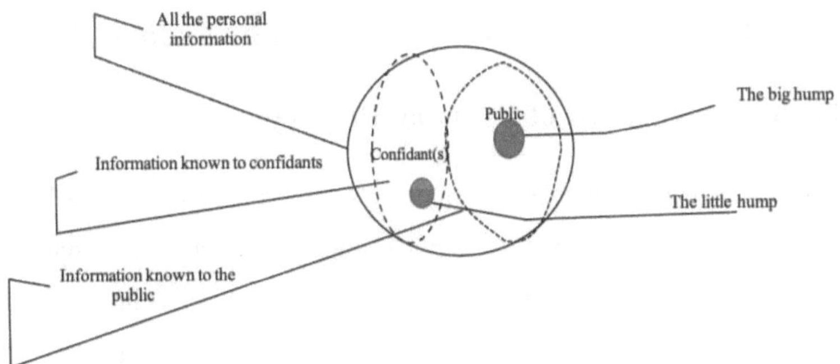

The relationship between types of personal information in the age of ICTs

---

[265] Ibid.

Deep personal privacy information (information known only to the person in question) is all the area outside the dashed and dotted ovals.

General personal privacy information is the entire area of the outermost oval minus the area of the dotted oval.

The relationship between the various types of personal information is changing due to the ICTs' increasing ability to expose personal information that was previously considered private. This refers both to the extent of confidants' knowledge of deep privacy information and to the general public's increasing ability to know pieces of privacy information – deep and general alike. The exposure of personal privacy information creates the "humps" – information about the individual that is known to others (one or more confidants or the general public) but not known to, possessed by or controlled by the individual who is the information subject. On the graph, the two smaller ovals inside the dotted and dashed ovals represent the humps.

Personal Privacy Information Carried by Each of the

Basic Components of Personal Privacy

Body: The body component includes physical characteristics such as appearance, height, weight, strength, health, senses, etc. Every one of those can be translated into information. Some physical information related to our body is stored in medical databases in the form of photographs or measurable values. Other elements of body-related information, such as appearance, age and gender, are stored in photographs or government databases such as population registers.[266] Some body-related information can only be stored and aggregated thanks to ICTs, like the results of personalized genetic testing, which has become possible after the completion of the Human Genome Project. The documented genetic data of millions of people are used for a variety of purposes, including the discovery of genetic variations that may cause diseases, the identification of criminals, and the testing of potential partners. Today, all the information related to our body and its properties is digitally recorded and aggregated. Whereas before the age of ICTs, most of that information would expire with the person, nowadays, it can survive long after the person's death and decay of the body.[267] Some elements of body-related information reside in the sphere of **general personal privacy** and are only shared with chosen confidants – e.g., our private parts, genetic information and other health information. Body-related information not known to anybody but ourselves resides in the sphere of **deep personal privacy**.

Mind: The abstract, virtual part of a human being that contains our thoughts, emotions, pains, desires, sense of self, memories and other mental elements such as inhibitions, fears, weaknesses and mental health conditions. All these elements turn to information when expressed through speech, writing, art (drawings, music, songs or dance), body movements or the positioning of the body. Until the age of ICTs, most of that information was short-lasting, with a little of it retained in analog formats such as writings, voice recordings or works of art. However,

---

[266] A person's health can be translated into various measurable factors such as blood pressure, heart rate and blood sugar.

[267] Digital systems forget nothing, and that raises questions regarding posthumous privacy – questions which had seemed irrelevant before the digital age.

these days, vast amounts of mind-related information are being aggregated in digital databases. Mental elements that had never been recorded until the age of ICTs include our culinary preferences, which are recorded in the databases of food retailers, and our hobbies and interests, inferred based on our shopping habits.[268] Human behavior is being recorded and turned into digitally stored and aggregated information.[269] A large portion of mind-related information is obtained from indirect sources. Later, I will elaborate on how ICTs (which reside in Popper's World 3) combined with databases (which reside in World 1) can uncover people's inner world, which is defined as deep privacy. The majority of mind-related information resides in the sphere of **deep personal privacy**; the parts we choose to share with confidants reside in the sphere of **general personal privacy**. I believe ICTs transfer mind-related information from the private sphere to the public sphere.

Actions: ICTs allow for constant monitoring of everything people do, both in the physical and virtual space. Most of our actions leave footprints in the physical space, which are collected and stored in large interconnected databases. Vast amounts of information are collected through our smartphone – when we are active (using apps), but also when we are not actively using it – and through a wide range of devices and sensors such as CCTV cameras (commonly used in "smart cities"[270]), private web cameras, geolocation satellites that can locate a

---

[268] Bernardo Periñán, "The Origin of Privacy as a Legal Value: A Reflection on Roman and English Law", *American Journal of Legal History*, vol. 52 (2) (April 2012), 185.

[269] Shoshana Zuboff, *The Age of Surveillance Capitalism* (London: Profile Books, 2019), 8.

[270] According to Margaret Rouse, "A smart city is a municipality that uses information and communication technologies to increase operational efficiency, share information with the public and improve

particular person in the physical space at any given moment, and "cookies"[271] planted in our PCs and smartphones to monitor our behavior in the virtual space. Until the age of ICTs, people's actions were hardly documented at all, but today, our movements in the physical space are recorded by apps like Waze and various proximity tracing apps,[272] our shopping habits are traced based on our credit card activity,[273] our interests can be found out based on what we order online and our surfing patterns (the websites we visit, the amount of time we spend on them, and our emotional responses to the content we see),[274] there are records of our typing behavior, phone and Internet communications,[275] eating habits,[276] movie preferences, participation in social events such as weddings or rallies, and many more. As we will see, most of those records are indirect information that is being used by government and

---

both the quality of government services and citizen welfare". https://www.techtarget.com/iotagenda/definition/smart-city (Accessed April 9, 2024)

[271] For more information on cookies, see Benoist, "Collecting Data for the Profiling...", 169-184.

[272] Digital proximity tracing and contact tracing apps came into public awareness during the COVID-19 pandemic as tools used by governments to prevent spread of the disease.

[273] Periñán, "The Origin of Privacy...", 185.

[274] Ibid., 186. The author argues this is a violation of privacy that one cannot protect oneself from because the actions in question are performed in the public sphere; to use his own expression – "It is like being naked in the village square". I disagree, and believe it to be a precise example of the kind of privacy invasion discussed in this book.

[275] Which can give a picture of our social contacts.

[276] By means of apps like DayTwo – a personalized nutrition app for people with type II diabetes developed by the Weizmann Institute of Science.

political organizations (such as intelligence and security agencies) as an aid for making political decisions or by commercial organizations as an aid for making commercial decisions.

Most of the things people do as agents who interact with the physical world are visible and observable in the public sphere; some things are shared exclusively with confidants and thus belong in the sphere of **general personal privacy**, while things only the person in question knows they did reside in the sphere of **deep personal privacy**, and as far as everybody else is concerned – might not have been done at all, or at least not by that person.[277] However, this changes when information about actions intended to be kept in **deep personal privacy** becomes known to others.

Property: The concept of property includes one's body and mind – of which one is the sole owner; the physical space in which one can do whatever one likes; and the products of one's mind, including abstract ones such as thoughts and memories. Property-related information has been recorded and stored since the dawn of civilization, especially since the invention of writing, but most of it would expire with the owner – i.e., most property-related information was inseparable from its container. Information related to intellectual property focused until the age of ICTs on copyrights and patent rights – products of World 2 manifested mainly in World 3. Privacy information such as feelings, intimate relationships, and body-related information was recognized as property long before the age of ICTs. This kind of property-related information belongs in the sphere of deep personal privacy, and until the age of ICTs, its recording had been minimal and limited to confidants.

---

[277] E.g., if it is known that a theft has been committed, but the identity of the thief is unknown.

ICTs allow such information to be recorded and digitally stored in interconnected online big data pools. For the first time in history, property-related information can be separated from its owner and get a life of its own.

External entities: Other players in the system of balances between the protection and violation of one's privacy. External entities include one's confidants and may be individuals or organizations (e.g., commercial companies, governments, or social networks such as Facebook and Google). External entities possess private personal information about one, including intimate information, information deemed confidential by social norms, and information whose confidentiality is protected by law, like doctor-patient confidentiality or attorney-client privilege. The massive sharing of private information on social media in the age of ICTs has greatly increased both the number of online entities that possess our private personal information and the amount of private information possessed by external entities[278] in a way that is unprecedented in its scope and in its impact on personal privacy and our ability to protect it.

Relationships with external entities: The nature of the relationships between an individual and other entities. Personal privacy information about personal relationships includes the intensity of each relationship (usually measured by the amount of time invested in the relationship, its emotional strength, level of intimacy and extent of mutuality[279]), the number of entities with which one communicates directly or indirectly, the methods of communication (e.g., email, phone

---

[278] Nearly every smartphone app records the physical location of the phone user, and I have already mentioned the constant and unstopping recording of our activities by IoT devices and social networks.

[279] E.g., knowing whether the other knows something about us.

or WhatsApp), the power relations between the parties, and the kind of privacy information communicated by one of both parties to the relationship. While before the age of ICTs, information about relationships with other entities – e.g., meetings with confidants – was usually documented only partly and in analog formats, with social and intimate relationships mostly not documented at all, nowadays most of that information recorded and stored in big data pools using the means we have already discussed. The information can be direct – e.g., images obtained from security cameras, "bugged" phone calls or intercepted emails, or indirect – such as the number of relationships, the type of information exchanged and its flow direction. Information about relationships with external entities belongs in the sphere of **general personal privacy** because it is also known to the other party (or parties) to the relationship. **External entities and one's relationships with them make up one's social network of privacy**.

Autonomy: The ability to control and make independent decisions concerning one's own body and mind without any external intervention. All three aspects of autonomy – personal, moral and political – are part of World 2, and therefore, any information regarding a person's autonomy can only be obtained indirectly: inferred based on that person's actions and overall behavior, on the behavior of external entities – which obviously affects the autonomy of the person in question, and on other indirect information collected about that person. Autonomy is undoubtedly an essential component of privacy, but despite that, most of the information related to it is out in the public sphere, with only some part of it residing in the sphere of **general personal privacy**.

Identity: Our identity is the product of our unique individuality, and we have the right to express it and live it out,[280] for even though every individual is made up of the same universal building blocks (such as common human physiology, genetics and cognitive abilities), the particular combination of those, together with every person's unique history, is what creates each individual's unique selfhood.[281] According to the liberal democratic approach, this special right must be protected. I will use Raffaele Rodogno's definitions of the various "identity questions" to define which information is carried by the identity component of privacy.[282]

(1) Passport identity: Carries basic bureaucratic information such as identification number, passport number and/or social security number. Until the age of ICTs, this information had been stored in analog format or in designated digital databases such as population registers and tax authority registers; today, it is replicated and disseminated among multiple interconnected databases. One can use passport identity information as a unique identification key to connect different databases together. We also have new biometric means of unique identification, such as face recognition, DNA and identification by IP (Internet protocol) address of a computer or mobile phone.

(2) Numerical identity: Identity based on one's cognitive abilities. Information about our cognitive abilities is recorded and stored in IQ tests and report cards from preschool to university. Before the age

---

[280] Rabi, *Omes ha-individualiyut*, 255.

[281] Garcia, *Individuality…*, 234.

[282] Rodogno, "Personal Identity Online", 309-328.
For a more detailed explanation of each "identity question", see Chapter 6. Here, I will focus on the elements of information carried by each type of identity.

of ICTs, it was stored in analog formats or in designated digital databases of schools and military systems; nowadays, it can be accessed in all those databases using passport identity information.

(3) Attribution identity: Before the age of ICTs, information related to attribution identity was mostly stored in databases controlled by official entities, such as population registers, tax authority registers, banks and social welfare institutions. Those databases were isolated and mostly static islands of information updated only in case of major life events like marriage, divorce, childbirth and death. In the age of ICTs, the monitoring of our actions, location, shopping habits and other patterns of behavior allows for the aggregation of all these data as information in interconnected big data pools.

(4) Social function identity: Until the age of ICTs, information about this type of identity had hardly been stored or aggregated. Small parts of it were stored in population registers or appeared in passport identity information (e.g., gender and ethnicity). Today, ICTs allow people to assume different identities in different applications, and sometimes more than one identity in the same application, and to present different faces depending on the place, time and other circumstances. On the other hand, the constant monitoring of people's activities by means of ICTs allows for aggregation of data in interconnected big data pools, and AI technologies are used to build profiles based on that data in many Internet-based applications. Advanced AI technologies combine different profiles into a single, nearly complete identity that can be associated with a specific individual, violating their personal privacy. There is an "arms race" of sorts between individuals' ability to assume multiple distinct identities and the technological capability to unify them, voiding the individual's efforts. Currently, the technological capability

to make a link between the same person's different identities seems to prevail over the human attempt to maintain multiple identities.[283]

(5) Attachment identity: This type of identity resides almost entirely in the sphere of deep personal privacy; any information about it can exist outside our body and mind only if we choose to share it – usually only with confidants. Later, I will show how ICTs can help expose parts of this information, violating people's personal privacy.

Anonymity: According to Pfitzmann and Hansen, the anonymity of a subject means that the subject is not identifiable within a set of subjects (the anonymity set).[284] As I have already mentioned, the capabilities of anonymity violation have greatly increased over time due to ICTs' ability to store, distribute, and connect between pieces of indirect information about people, eventually allowing their identification. Anonymity violation has two stages: at the first stage, a person's identity is revealed; at the second stage, that identity is associated with a specific individual to physically get to them. I will use Michael Birnhack's definitions of different types of anonymity to describe the information carried by each type and the players engaged in it.[285] **One thing that is common to all types of anonymity is the sense of anonymity as anonymity of information.** Birnhack lists several categories of social anonymity:

Anonymity of donation: In charity donations, the donor and the recipient, and sometimes both, may remain anonymous. Sometimes donations are mediated by charity organizations, usually as a way to increase anonymity. The same type of anonymity applies to welfare payments, in which the mediator is the government, the "donor" is

---

[283] Schechtman, "Stories, Lives, and Basic Survival...", 155-178.
[284] Hansen and Pfitzmann, "A terminology for talking about privacy...".
[285] Birnhack, *Merkhav Prati: Ha-zkhut la-pratiyut ha-ishit...*, 269-286.

society as a whole – which remains anonymous because no specific welfare payment can be associated with any specific taxpayer, and the recipient is anonymous because social norms dictate that welfare payments should be confidential.

Anonymity of bodily donations: Including gamete, organ, bone marrow and blood donations. The donor usually remains anonymous, and sometimes the recipient does as well. The donation is almost always mediated by a medical institution to ensure physical compatibility. The same institution usually also performs the medical procedures required for the donation.

The anonymity of exams in academic institutions: According to the academic norm of anonymous exams, student exams are checked and graded without knowledge of the student's identity. The anonymity is one-sided, as the professors who grade the exams are not anonymous. The institution office is a confidant who knows both parties.

Anonymity of peer review: Applied in academic publications and scientific journals. The anonymity is mutual: the authors and the reviewers are anonymous to each other. The process is always mediated by a bureaucratic entity that knows the identities of both parties.

The anonymity of tenders: Bids are submitted anonymously to ensure the entire decision-making process is free from bias, conflicts of interest and other irrelevant deliberations. The process and the identities of the tender committee are not anonymous.

Anonymity of medical tests: The anonymity of the patient is maintained in medical tests, especially tests for diseases for which one may be subject to discrimination (e.g., at work), shaming or social stigma. The anonymity applies both to the fact of the test performance and to the identity of the patient. In the age of ICTs, this type of anonymity is at great risk because medical data is aggregated and stored in interconnected databases, which makes it very easy to associate a test with a specific patient.[286]

Anonymity in therapy groups: Therapy groups such as Alcoholics Anonymous maintain the anonymity of their members from anybody who is not a member of the group. The support group becomes a confidant in this case.

Commercial and consumer anonymity achieved by using cash: The majority of commercial transactions that involve an exchange of money or goods are documented. Documentation of trade relations began during the Industrial Revolution and the rise of nation-states. According to Frank Webster and Kevin Robins, scientific management of industrial, commercial and economic systems (Taylorism) is impossible without an information system such as double-entry bookkeeping. However, this compromises consumer anonymity by extending the circle of confidants to include banks, tax authorities and other regulatory agencies, who document and store each party's passport identity at the very least.[287] Before the age of ICTs, people would pay in cash to maintain their consumer anonymity and could thus avoid identification; however, nowadays, commercial and consumer anonymity is increasingly violated, and Internet companies and social media exploit it to maximize their profits. Recent government regulations aimed to battle tax evasions from black market deals greatly limit the use of cash. On the other hand, the advent of virtual currencies allows for anonymous transactions.

Anonymity of communications: ICTs have had a major impact on our ability to communicate anonymously. The ICT properties

---

[286] I will discuss it in further detail in the section on *k*-anonymity.

[287] Robins and Webster, *Times of the Technoculture*, 102-109.

described in Chapter 1 make anonymous communication virtually impossible. Birnhack focuses on the anonymity of **personal communications**, which is applied both to the fact of communication and its content. In this kind of communication, the parties are each other's confidants.[288] Until the age of ICTs, anonymous communication could be face-to-face, by phone (free from third-party listeners), or by post, where the anonymity of the content was protected by a sealed envelope. As already mentioned, this has completely changed in the age of ICTs. The anonymity of customer-vendor communications has also become impossible to maintain because most of them are mutually recorded. Above all that, the characteristics of our communications cross-referenced with metadata, such as our web surfing patterns, consumption habits and geographical location, can be used to create individual users' profiles with the aid of AI tools, leading to violation of all the various types of anonymity.

Anonymity in legal procedures: Publication of suspect names and of the very facts of a legal proceeding can be prohibited. A trial may be conducted behind closed doors without any public disclosure of the details of the case and the legal process. The parties involved, their attorneys, the courts and law enforcement agencies serve as confidants. Anonymity may be applied to the suspect's or defendant's identity and/or to the details of the offense.

Civic/political anonymity: The right to confidentiality of political opinion and to secret elections. In the age of ICTs, this type of anonymity is being challenged by the use of AI technologies on cross-referenced data from big data pools.[289]

---

[288] Birnhack, *Merkhav Prati: Ha-zkhut la-pratiyut ha-ishit…*, 284-285.
[289] See, for example, the Facebook–Cambridge Analytica data scandal.

Civil anonymity: The right and ability to make an anonymous report to law enforcement agencies, anonymous whistleblowing, and reporter source confidentiality. The whistleblower or source remains anonymous to those they are reporting against and has confidants. The reported act and the people who supposedly committed it are not anonymous.

Anonymity of expression: Applicable mainly to web and social media activity, including "active" activities such as blogging, posting and responding to other people's posts, as well as "passive" activities such as web surfing. It is clear why people may want to remain anonymous when using "delicate" web services such as dating and pornographic sites, but some also choose to do their online shopping anonymously so as not to reveal their consumption preferences. Using a pseudonym online allows people to express themselves freely and say whatever they think without filters (whether this is necessarily a good thing is a different question) and with minimal social and other risks. In their strife for online anonymity, people create online identities that seem to have a life of their own. Thus, on the one hand, the age of ICTs has opened up new opportunities for anonymity, but on the other hand, the same ICTs constantly violate that anonymity. There is an "arms race" of sorts between individuals' desire to maintain and protect their anonymity and others' desire to violate it, and right now, anonymity seems to be losing.

Anonymity, in its broadest sense, includes not only one's identity but also one's social connections, activities, etc. All the basic components of personal privacy – and, accordingly, the information they carry – deserve the amount of anonymity that is accepted in the culture in question.

The following table shows the parties involved in the preservation or violation of anonymity:

Type of anonymity	Type of party						
	Anonymous party	Order I confidant	Order II confidant	External parties	Anonymity violation I	Anonymity violation II	Anonymity violation III
Social anonymity	Donation recipient	Donor	Mediator between donor and recipient	Everybody except the recipient, the donor and the mediator	Exposure of the recipient[1]	Exposure of the donor	Disclosure of the fact a donation has been made or received
Equality-ensuring anonymity	Student	Teacher/examiner	Academic office/admission committee	Everybody except the students and confidants	Exposure of the student's identity to the examiner		The fact that the test is not anonymous.
Medical anonymity	Patient	Healthcare provider	Healthcare organization/hospital	Everybody except the patient and confidants	Exposure of the patient's identity to other people		Disclosure of the performance of the test
Commercial anonymity	Customer or merchant	The other party to the transaction, whether individual or organization	Regulatory agencies, e.g., tax authorities	Everybody else	Exposure of the customer	Exposure of the merchant	Disclosure of the facts and details of the transaction
Anonymity of communications	Party A	Party B		Everybody else	Exposure of Party A	Exposure of Party B	Disclosure of the facts of communication or its contents
Legal anonymity	Suspect or defendant	Prosecutor or plaintiff	Attorneys, courts, law enforcement agencies	The public	Exposure of the defendant's identity and other info		Disclosure of the details of the case
Civil/political anonymity	Whistleblower	Reporter/law enforcement agencies		The public	Whistleblower identification and association with report		
Anonymity of expression	Person active online		Internet companies		Identification of the person		Disclosure of online activity and its nature

1. Exposure – Disclosure of the information to anyone except the donor and confidants.

Contextual Integrity

In her book *Privacy in Context: Technology, Policy, and the Integrity of Social Life*, Helen Nissenbaum looks at privacy as a flow of information.[290] On one end is the information sender, who shares personal privacy information; on the other end is the information recipient. The answer to the question of whether the transmitted information should be considered personal privacy information depends on the social and cultural context and the norms that govern the various realms of social life. Information considered private in one culture may be considered personal but not private in another. Thus, according to Nissenbaum's contextual integrity theory, privacy is culture-dependent. This theory supports the previously introduced argument that there is a mutual influence between culture and the definition of personal privacy and adds the important nuance that personal privacy information is also culture-dependent.

The following table shows the associations between the categories of personal privacy information and the six conceptions of privacy described by Solove.

A "V" in a cell means the conception in the corresponding column includes the information category in the corresponding row.

---

[290] Nissenbaum, *Privacy in Context...*, 127.

		The right to be let alone	Limited access to the self	Secrecy	Control over personal information	Personhood	Intimacy	Information obtained directly or indirectly	Info category No.
		Solove's conceptions of privacy							
Information carried by the basic components of privacy	Private space information	V	V				V	Directly	1
	Body-related information	V	V			V	V	Both	2
	Mind-related information	V	V			V	V	Indirectly	3
	Information about actions	V	V	V	V		V	Both	4
	Property information	V			V	V		Directly	5
	Information about external entities	V	V	V	V		V	Both	6
	Information about relationships with external entities	V	V	V	V	V	V	Both	7
	Information about the values that guide decisions and actions (autonomy information)		V	V	V	V	V	Indirectly	8
	Anonymity information	V	V	V	V	V		Both	9
	Identity-related information			V	V	V		Both	10

The personal privacy information categories could be broken down further and summed upwards, but for the sake of simplicity and generality, I will keep it at this level.

Personal Privacy Information Qualifies as "Information"

Privacy information meets the criteria that define an entity as information. Because the word "information" has many different meanings and definitions, I will here use Luciano Floridi's general definition of information (GDI).[291] According to Floridi, entity α is an instance of information if and only if:[292]

1. It contains N data ($N \geq 1$);
2. The data are well-formed;
3. The data are meaningful.

The first criterion is that information should be made up of data, i.e., it cannot be dataless. The absence of a particular datum does not cancel existing data but rather provides additional information.[293] Anonymity, for example, is the absence of data. In the previous sections,

---

[291] Luciano Floridi, "Philosophical Conceptions of Information", in: *Formal Theories of Information: From Shannon to Semantic Information Theory and General Concepts of Information*, ed. Giovanni Sommaruga (Springer, 2009), 13-53.

[292] We will see how all privacy information meets these criteria.

[293] For example, if I do not know the cause of a person's death, it does not cancel my knowledge of the fact they died, or the existence of a cause of death in general. Rather, it means a cause of death exists, but is currently unknown.

I described the association between the basic components of privacy and privacy information, the categories of personal privacy data, and the way ICTs collect and store that data in big data pools. Thus, personal privacy information meets the first criterion.

The second criterion requires that the data be well-formed, i.e., organized according to the rules (syntax) that govern the chosen system. The syntax here must be understood in its broadest sense as what determines the form, construction, composition or structuring of something. It may mean code, musical notes or signs on a mechanical drawing, and not necessarily the syntax of a language. The paradigm of personal privacy, which frames all the main conceptions of personal privacy (see Chapter 7), includes symbolic generalizations, which are the syntax according to which privacy information is organized. Thus, personal privacy information also meets the second criterion.

The third criterion is that the data should be meaningful, i.e., comply with the semantics of the chosen system. Floridi makes a distinction between the information's meaningfulness and our ability to understand its meaning.[294] The values and metaphysical parts of the paradigm of personal privacy give privacy information meaning in terms of protection and violation of personal privacy; we can see how the six conceptions of personal privacy, on the one hand, include the various categories of personal privacy information, and on the other hand, represent the metaphysical parts of the paradigm of personal privacy. Thus, personal privacy information also meets the third criterion of the GDI.

---

[294] Floridi, "Philosophical Conceptions of Information", 22.

Privacy Information is a Well-Defined, Objectively Valid Concept

In Chapter 5, I described the difficulty of discussing personal privacy due to the lack of a clear and extensive definition thereof. To overcome this difficulty, I am proposing to replace the discussion of the influence of ICTs on personal privacy with a discussion of the influence of ICTs on personal privacy information – a concept that is much clearer and more measurable.

Let us now see how each category of personal privacy information can be associated with one or more of the five categories of data defined by The Stanford Encyclopedia of Philosophy.[295]

(1) Primary data: Usually obtained through indicators; e.g., high temperature is an indicator of illness. The categories of personal privacy information that contain primary data are body-related information, information about actions, information about mental and psychological factors, and identity-related information.

(2) Secondary data: Data whose absence is what provides information and meaning. Anonymity information is the category that contains most secondary data, as its very nature is the absence of identifying information. The absence of data can also provide information about relationships with external entities, actions (in particular – inaction), decisions and values.

(3) Metadata (data about data): In the context of personal privacy information, metadata includes information that defines one's

---

[295] Stanford Encyclopedia of Philosophy, "Semantic Conceptions of Information", first published Wed 5 Oct, 2005; substantive revision 14 Jan, 2022.

identity, for example, information about interactions with other people that define one as selfish or one's actions and decisions that define one as a criminal. Because metadata are a subtype of personal privacy information and do not belong in any specific information category, they are not included in the summary table below, but they do **qualify as personal privacy information**.

(4) Operational data: Data that have operational use. Include private space information, passport identity information, and information about the values that guide one's decisions and actions (autonomy information).

(5) Derivative data: Data that derive and can be logically inferred from other data. This includes indirectly obtained mind-related information, body-related information inferred from physical indicators (e.g., presence of illness inferred from high temperature), information about actions or decisions inferred based on their observable consequences, indirect information about relationships with external entities, and identity-related information revealed based on other data.

The following table summarizes the classification of personal privacy information by category of data as defined by The Stanford Encyclopedia of Philosophy. A "V" in a cell means that the information category in the corresponding row includes the data category in the corresponding column.

		Category of data			
		Primary data	Secondary data	Operational data	Derivative data
Category of personal privacy information	Private space information			V	
	Body-related information	V			V
	Mind-related information	V			V
	Information about actions	V	V		V
	Property information	V		V	
	Information about external entities			V	
	Information about relationships with external entities		V		V
	Information about the values that guide decisions and actions (autonomy information)		V	V	V
	Anonymity information		V		
	Identity-related information	V		V	V

As you can see, personal privacy information categories can be aligned with the Stanford Encyclopedia of Philosophy definition of data categories. This means personal privacy information is a well-defined concept.

If we try to answer the following ontological questions regarding privacy information, we will see that all the various types and categories of personal privacy information can, with the aid of ICTs, be placed in Popper's World 3. The entities that inhabit World 3 have an objective status, so now, when we talk about deep privacy, we can replace the discussion of subjective entities such as feelings, experiences and desires with a discussion of information about those entities, which is an objective entity as far as world 3 entities are objective. The ontological questions I would like to ask regarding privacy information are as follows:

Can personal privacy information be detached from its physical vessel, i.e., the individual? Until the age of ICTs, personal privacy information had normally been attached to and inseparable from the individual[296] and expired with them (unless the individual was a public figure, in which case parts of their personal privacy information were preserved in history books). However, in the age of ICTs, both general personal privacy information and growing parts of deep personal privacy information are stored in data pools, detached from physical persons. What makes this detachment possible is the ability to translate any category of personal privacy information into bits of 0 and 1. Once detached, the information – in whole or in parts – can be digitally stored across different databases and disseminated through the Internet. In other words, ICTs are allowing us to place any element of personal privacy information in world 3 and to connect it with other entities that inhabit that world, such as information analytics or knowledge in psychology

---

[296] With the exception of passport information, medical information etc.

and social sciences. All these tools help make connections between raw data and metadata, giving them meaning. Moreover, World 3 entities help associate data with operational tools and derive new data from given information, e.g., derive your preferences based on your shopping data. By consolidating those data, companies can infer market trends based on the preferences of several individual people.

Can John Archibald Wheeler's "it from bit" doctrine, which says that "all things physical are information-theoretic in origin,"[297] i.e., that every physical item is at its bottom an information bit that has two possible states – 0 or 1 (true or false), be applied to personal privacy? In other words, can the discussion of personal privacy be replaced with a discussion of personal privacy information? The answer is yes, firstly, because each of the basic components of personal privacy carries elements of personal privacy information, which means personal privacy information covers every aspect of personal privacy, and secondly, because personal privacy information meets the GDI requirements for being information and is well-defined based on The Stanford Encyclopedia of Philosophy.

Is information the name of the content that we exchange with the outside world? In our case, personal privacy information is the content of any discussion of personal privacy. For example, the philosophical discussion of the right to be let alone[298] can be regarded as a discussion of one's right to limit the disclosure of one's personal privacy information, for as Birnhack says in his book, monitoring of

---

[297] John Archibald Wheeler, "Information, Physics, Quantum: The Search for Links", in: *Complexity, Entropy, and the Physics of Information*, ed. Wojciech H. Zurek (CA: Westview Press, 1990), 313.
[298] Warren and Brandeis, "The Right to Privacy".

individuals by means of ICTs gnaws at our privacy,[299] and what is that "monitoring" but the transfer of personal privacy information from the private sphere to the public sphere? Thus, replacing the discussion of personal privacy with the discussion of personal privacy information is justifiable.

Gregory Bateson defined the elementary unit of information as "a difference that makes a difference."[300] In the case of personal privacy, it may mean, for example, the ability to differentiate between people based on different body-related information.

The above ontology of personal privacy information shows that it is an adequate representation of personal privacy because:

Based on Question 1: Privacy information can be digitized and detached from the individual who is the information subject;

Based on Question 2: Every physical thing is at its bottom a bit of information, and therefore, personal privacy is also essentially personal privacy information;

Based on Question 3: Any discussion of personal privacy is, in fact, a discussion of personal privacy information;

Based on Question 4: The personal privacy of individual X is differentiated from the personal privacy of individual Y, at least by their respective privacy information.

These four arguments lead to the conclusion that any discussion of personal privacy can be replaced with a discussion of personal privacy information. The ontology of personal privacy information that we have just seen clearly shows that there is no logical obstacle to such a replacement or, in other words, to a numeric

---

[299] Birnhack, *Merkhav Prati: Ha-zkhut la-pratiyut ha-ishit...*, 35.
[300] Gregory Bateson, *Steps to an Ecology of Mind* (New York: Ballantine Books, 1972), 428.

representation of personal privacy information. The digitization of privacy information is the very thing that allows ICTs to store, process, analyze and disseminate personal privacy information. Thus, not only is it a well-defined concept, but it is also measurable and quantifiable in accordance with Claude Shannon's information theory.[301]

Some personal privacy information is environmental information, formally defined as "two systems A and B related so that the presence of A as a category or state F is associated (correlates) with B, and the presence of B as a category or state G can be inferred based on information." In the context of personal privacy, a good example of environmental information is the discovery of one's identity by one's fingerprint. AI technologies make extensive use of environmental information – i.e., signs that are observable in the physical world – to uncover personal privacy information, general and deep alike. One assumption behind such use of AI technologies on environmental information is that there is a valid match between physical items and signs observable in the physical world – i.e., Popper's World 1 – and what is going on in World 2. Another assumption is that we can make valid conclusions about World 2 based on information inferred by AI technologies using statistical adjustment and generalization rules. I am not saying that those adjusted and generalized conclusions are absolute truth, but I will say that the information and knowledge obtained through adjustment and generalization do apply to specific individual people at a given time and place.

---

[301] Shannon, "A Mathematical Theory of Communication", 12.

Chapter 9

Personal Privacy in the Age of ICTs

The Information Society – The Network Society

Around the end of the second millennium, a number of major social, technological, economic, and cultural transformations came together to give rise to a new form of society – the network society.[302]

According to David Lyon, the roots of the ICT age were planted in the transition from a Fordist society based on the manufacturing industry to a post-Fordist society based on service and knowledge industries.[303] Daniel Bell says post-industrial society is moving toward an industry centered around science and skilled tech professionals who possess what he calls "theoretical knowledge."[304] According to Eran Fisher, the shift to an industry based on information and knowledge constitutes a change in the industrial-technological paradigm from a centralized, hierarchical system to a decentralized network system.[305]

---

[302] Manuel Castells, *The Rise of the Network Society*, Preface to the 2010 edition, xvii.

[303] David Lyon, "From 'Post-Industrialism' to 'Information Society': A New Social Transformation?", *Sociology* 20 No. 4 (November 1986), 577-588.

[304] Daniel Bell, *The Coming of Postindustrial Society: A Venture in Social Forecasting* (New York: Basic Books, 1973), 14.

[305] Eran Fisher, "Lekhudim ba-reshet: siakh ha-technologiya ha-rishtit ve ha-capitalism ha-khadash" [Caught in the net: the new capitalism and network technology discourse"] (in Hebrew), *Teoriya ve Bikoret* 37 (Fall 2010), 154-156.

What enabled this transition from Fordist to post-Fordist capitalism, which began – according to most scholars – sometime in the early 1970s, was the development of ERP[306] and CRM[307] information systems. Without those systems, the transition would not be possible, and that means information has been its most essential resource from the very beginning.

Post-Fordism evolved into a network society. In Chapter 1, we saw Toffler's, Castells' and Floridi's predictions that in the age of ICTs, information would be the primary resource around which the entire society would be organized; in this chapter, we will see how those predictions came true. All three scholars defined the age of ICTs as a revolution. Castells points out the paradox between the way ICTs increase social integration and the way they undermine the traditional Western conceptions of individuality and independence.[308] This paradox results from the impact of the ICT revolution in the field of personal privacy. Therefore, I will describe the age of ICTs from the "network" aspect, recognizing its major importance as a key element of this age. The digitalization of the physical and mental worlds (Popper's Worlds 1 and 2) is an integral part of network technology. Apparently, it is not an imposition of some mathematical technological whim upon the world but a revelation of the natural order of things.[309] Based on the digital discourse, the enormous power to generate social change possessed by

---

[306] Enterprise resource planning system: a data processing system for integrated management of all the main business processes, such as human resources, finances, marketing, logistics and operations.

[307] Customer relationship management.

[308] Castells, *The Rise of the Network Society*, 5.

[309] Eran Fisher, *Capitalism be-idan ha-tikshoret ha-digitalit* [Capitalism in the Age of Digital Communications] (in Hebrew) (Tel Aviv: Resling, 2011), 233.

network technology more than by any previous technological paradigm stems not only from its being more natural but also from our culture becoming more technological.[310] According to Yochai Benkler, "Enabled by technological change, we are beginning to see a series of economic, social, and cultural adaptations that make possible a radical transformation of how we make the information environment we occupy as autonomous individuals, citizens and members of cultural and social groups."[311] ICTs' increasing presence in all the Three Worlds defined by Popper is reflected in the words of major Internet corporation executives: former Google CEO Eric Schmidt was quoted saying, "We need the information about yourself and your friends to make our products work better," and "We know where you are. We know where you've been. We can more or less know what you're thinking about",[312] while Microsoft CEO Satya Nadella said, "Microsoft understands all physical and institutional spaces, people, and social relationships as indexable and searchable."[313]

Lastly, technologies, even in the narrowest sense of tools and machinery, are a representation of knowledge. They are constituted as phenotypes of both knowledge and power and cannot be analyzed in terms of only one or the other. In Castells' words, "Technology is society, and society cannot be understood or represented without its technological tools."[314] In view of all the above, I will now describe the ways personal privacy is influenced and reshaped by network technology

---

[310] Ibid., 244.

[311] Yochai Benkler, *The Wealth of Networks: How Social Production Transforms Markets and Freedom* (New Haven and London: Yale University Press, 2006), 1.

[312] Zuboff, *The Age of Surveillance Capitalism*, 498.

[313] Ibid.

[314] Castells, *The Rise of the Network Society*, 5.

and discuss those influences from the perspectives of four components of network economy: network production, network work, the network market, and the network human.

Network Production and Services

One of the key characteristics of the network production paradigm is the replacement of large factories accumulated in one physical space and responsible for nearly the entire production process, from the raw materials to the final product, with smaller production facilities that specialize in a limited number of core activities and outsource the rest to vendors and subcontractors, resulting in a supply chain – or rather, a supply network – which is often spread across different continents, with customer-vendor relations between each pair of nodes. This decentralization of the production process often comes hand in hand with a methodological shift from bulk mass production in large standardized batches to mass customization,[315] with shorter production cycles, smaller batches and greater adaptability to the market's changing demands.[316] These developments have led to a paradigm shift from vertical organizations (hierarchies) to "flat" of "horizontal" organizations,[317] or what Castells calls "network enterprises" – a business structure

---

[315] Maya E. Dollarhide, Mass Customization https://www.investopedia.com/terms/m/masscustomization.asp (Accessed 14/1/2021): "Mass customization is the process of delivering market goods and services that are modified to satisfy a specific customer's needs."
[316] Robins and Webster, *Times of Technoculture*, 150-151.
[317] Fisher, *Capitalism be-idan ha-tikshoret ha-digitalit*, 127.

characterized by decentralized production and reduced middle management.[318]

The network production and service model requires that vendors and customers have intimate knowledge of each other so each customer can choose the most appropriate vendor (supplier or service provider) out of several options. The choice is made based on tenders or ROI (request of information) documents, in which the vendor is required to reveal much information about itself. Network production tends to follow the "long tail" strategy, which means offering a wide selection of less popular items rather than targeting the mainstream.[319] According to Chris Anderson, who coined the term "long tail," advanced modern-day technologies significantly reduce production costs for items like CDs, clothes, printed books and other products that can be sold online,[320] while online marketing cuts marketing and distribution costs and makes such products universally available; consequently, it is becoming more reasonable to produce niche products. The customization of production and services has given rise to prosumers – human agents who are, at the same time, producers and consumers of a particular product or service.[321] Some of the major examples given by Toffler include do-it-yourself furniture,[322] power tools for home use, and home pregnancy tests. The most prominent technological example of this producer-consumer merger is the 3D printer – a device that allows prosumers to self-produce

---

[318] Castells, *The Rise of the Network Society*, 163.

[319] Chris Anderson, *The Long Tail: Why the Future of Business Is Selling Less of More* (United States: Hyperion, 2006).

[320] For a more detailed discussion of this topic, including examination of case studies such as Amazon and Goggle, see the section on the network market.

[321] Toffler, *The Third Wave*, 282-305.

[322] IKEA is a good example of that.

a growing selection of items in their own private space.[323] However, the prosumption goes beyond the do-it-yourself industry. In a variety of manufacturing industries, the consumer can take part in the finalization of the product: some fashion retailers allow online shoppers to virtually "try on" pieces of clothing using augmented reality, while some car manufacturers allow customers to choose the car's design.[324] Healthcare is also becoming more customized with self-care and remote diagnostic tools.[325]

Another major example of prosumption is content creation on social media such as Facebook, Instagram, TikTok and Wikipedia, "allowing private users to collaborate and work together as a hive. In network production, what seems to be a mere act of consumption also modifies and optimizes the product. For example, Google translates the traffic and connectivity patterns formed by the two billion searches performed every month into the intelligence behind a new form of economics. Another example is Wikipedia, the collaborative online

---

[323] Three-dimensional (3D) printing, or additive manufacturing, is a technology that allows one to create a physical three-dimensional model directly from a CAD (computer-aided design) software. Most 3D printers use various polymers to print the model (prototype) layer-by-layer.

[324] Fisher, *Capitalism be-idan ha-tikshoret ha-digitalit*, 139-140; Shakhar Kissinger, "Al ofna, kniyot ve smartphone – kol aplicatziyot ha-ofna she yeshadregu lakhem et khavayat ha-kniya" [Fashion, shopping and smartphones – all the fashion apps that will upgrade your shopping experience] (in Hebrew) https://blog.partner.co.il/life/ (Accessed 10/2/2021)

[325] Tyto is a smart device with which patients can perform throat, ear, lung and other examinations from home and communicate with their physicians to receive remote diagnostics. Several healthcare organizations in Israel have begun to offer this service as part of their health plans.

encyclopedia: every time we make a connection between words, we teach it a new concept."[326]

Network production consists of two layers of networks. The first layer is a physical network comprised of nodes that exchange physical goods, while the second layer is an information network in which the nodes exchange information of different types, such as descriptions of the goods trafficked through the network (size, weight, price, expiration date, etc.) and commercial information that describes the relationships between the nodes. Because prosumers are nodes in the supply network, information about them flows through the network, at least information related to their identity. Some of that information is private. Customized production and services require the producer to have intimate knowledge of the consumer: for example, in the fashion industry, the producer should know the consumer's body size, material preferences (e.g., shape, color, type of fabric etc.), and more. The following table describes the elements of privacy information that flow through the production network by the previously defined privacy information categories:

---

[326] Fisher, *Capitalism be-idan ha-tikshoret ha-digitalit*, 142.

Information category	Information elements
Body-related information	Body size (height, weight) and body shape
Mind-related information	Shopping preferences
Information about actions	Items one has purchased in the past (in the next section, we will see how past purchases can be used to predict future ones)
Property information	Items one already owns (marketing practices such as offering a discount on a second item)
Information about external entities	Where and under what circumstances the transaction was made
Information about the values that guide decisions and actions (autonomy information)	For example, if one avoids buying products that contain animal ingredients
Identity-related information	One's passport identity, at the very least, must be known in order to make a transaction.
Anonymity	Impossible in this kind of online transactions

Elements of privacy information that flow in personalized medicine:

Information category	Information elements
Body-related information	Genetic traits, medical history, allergies and sensitivities
Mind-related information	Mental and psychological traits
Information about actions	Various lifestyle habits
Private space information	The environment one lives in, its climate and potential health hazards
Information about relationships with external entities	One's human environment and its influence on one's health
Information about the values that guide decisions and actions (autonomy information)	For example, values dictating eating habits (such as vegetarianism or veganism) may affect one's health.
Identity-related information	One's passport identity, at the very least, must be known in order to make a transaction.
Anonymity	Impossible in this kind of online transactions

Personal information (including private information) of prosumers that flows through the production network:

Information category	Information elements
Body-related information	Elements of information required of one as a specific customer or patient
Mind-related information	Information about oneself that one uses when writing posts, blogging or contributing to an online encyclopedia
Information about actions	Actions and lifestyle habits characteristic of a specific prosumer.
Property information	Place of residence, means of communication, etc.
Information about relationships with external entities	What other entities in the supply network one interacts with.
Information about the values that guide decisions and actions (autonomy information)	Values and norms that affect one's actions, decisions or feelings.
Identity-related information	Identifying information one is required to provide in order to be a node in the supply network.
Anonymity	Impossible for a node in the supply network.

According to Fisher, the network production methodology challenges the traditional concepts of production and consumption, allowing simple modification, adjustment and reuse of previous contents.[327] As a result, consumers are easily "upgraded" into producers. Thus, network production is not merely post-capitalist because it is blurring the production-consumption and producer-consumer dichotomy; it is also post-proletarian and post-consumer. It allows creativity, deep interest, personal expression and collaboration to penetrate and refresh the spheres of production and consumption alike. Individuals are no longer regarded as pawns in a mass consumption society; they become independent agents and prosumers. According to the digital discourse, network production has changed the capitalist balance of power between the industry managers and their representatives and the common workers. While Taylorist line laborers are just anonymous "screws in the system," network production allows workers to take part in the management of the production process and bring in their individual skills and creative abilities. The production network does not only encourage close, intimate relationships between producers and consumers; it cannot exist without them. The replacement of the assembly line with the production network and the resulting increase in the number of production nodes located in the consumer's private space is also blurring the dichotomy between the private and public spheres.[328] **A node in the production network can be in the private sphere and in the public sphere interchangeably – and sometimes simultaneously – without changing its physical location.**

---

[327] Fisher, *Capitalism be-idan ha-tikshoret ha-digitalit*, 139-140.

[328] Work from home and the use of Zoom, which have been on the rise since the COVID-19 pandemic, bring the public sphere into the private space, and vice versa.

The blur of the private-public dichotomy includes the physical merging of private and public spaces, the mixing of time devoted to private affairs and time devoted to work, and increased amounts of private personal information that flows between the nodes of the production network. That information flow, in particular, has a major influence on personal privacy as part of the general influence of the age of ICTs on personal privacy.

Network Work and the Network Worker

The advent of network production goes hand in hand with the advent of the network worker. The "networkization" of employment in the age of ICTs has three main characteristics:[329] a blur of the distinction between work and leisure, workspace and living space, work time and personal time; a blur of the distinction between managers (capitalists) and workers; and the growing importance of skill and professionalism. Characteristics 1 and 3 have a major impact on personal privacy. For the information worker, network technologies are not merely an instrument of work but an important factor of self-identity and social status. They are the digerati (a portmanteau of "digital" and "literati") – the new elite of the computer industry, social media and online communities, who carry what Castells calls "the spirit of informationalism," or "the spirit of the network."[330] This is an entirely new category of workers directly or indirectly involved with information and communication technologies, sometimes at the very forefront of the development, financing, operation and dissemination of new ICTs. The digerati are

---

[329] Fisher, *Capitalism be-idan ha-tikshoret ha-digitalit*, 103.
[330] Ibid., 104.

nodes in the production network, suppliers of the workforce and a new type of growth driver.[331] Unlike traditional Fordist employees, their time and location are flexible and not bound to the production process, so the time they devote to production work mixes with their leisure (private) time. According to Fisher, network technology is blurring the distinction between work and non-work, productive time and free time, office space and home space, which means work can be done anywhere.[332] The digerati are a manifestation of the profound transformation from Fordist to the post-Fordist conception of employment. The distinction between work and leisure is blurred, and as work becomes detached from a physical location, any space can be a workspace. One can work outdoors while enjoying nature or stay up to date with the company affairs while on vacation. Thus, the blurring of the distinction between office space and "outside" space may also mean greater binding of one's personal time to work. The blurring occurs both in terms of space, with much of the work being done from home (in Fordist culture – part of the private sphere), and in terms of time – with no clear dichotomy between work time and time devoted to personal affairs, for when you are working from home, work inevitably mixes with other things such as looking after your children, leisure activities, and visiting work-unrelated websites. This trend of mixing work with leisure and private space with public space was accelerated by the COVID-19 pandemic.[333] Moreover, the network

---

[331] Ibid.

[332] Ibid., 104-105, 108-109.

[333] Neil Corney, CEO of Citi Israel and high-tech executive, said in an interview from April 2020: "We are in no hurry to bring our employees back to the offices. The pandemic has overturned all our assumptions: while up until last month we had been fearing that working from home would compromise productivity, now we can say with absolute confidence that our employees are very productive from home."

industry has created a variety of jobs that can initially be done from home. Mary L. Gray and Siddharth Suri describe what they call "ghost work" – invisible "last mile" Internet jobs such as cleaning out offensive tweets and Facebook posts missed by AI applications.[334] Those ghost workers are employed on an "as needed" basis, i.e., for a specific project or task. According to Gray and Suri, the sector of human-AI integrated jobs is on the rise: in 2016, it included 20 million workers working from home part-time on an "as needed" basis, and by 2025, its turnover is expected to reach 2.7 trillion dollars.

Digerati no longer feel committed to the "old school" ethics, which demand a separation between work and play, work time and leisure time:[335] they may find themselves working – and can potentially work – anywhere, anytime. The coffee shop is penetrating the workplace, but the workplace is also penetrating the coffee shop. Paradoxically, network workers devote more of their personal time to work than Fordist workers. The decentralized work environment, on the one hand, increases workers' autonomy and individual expression, but on the other hand, it invades their private sphere more extensively. The new productive force is based on the technological ability to harness workers' time and skills to the production process in ways that were unknown and unused before the age of ICTs. In addition, network relationships are more flexible and can be made and unmade more easily. Workers can be recruited on a temporary basis for a specific project without forming

---

https://www.calcalist.co.il/internet/articles/0,7340,L-3808370,00.html (in Hebrew) (Accessed 9/2/2021)

[334] Mary L. Gray and Siddharth Suri, *Ghost Work: How to Stop Silicon Valley from Building a New Global Underclass* (New York: Houghton Mifflin Harcourt, 2019), 7.

[335] Fisher, *Capitalism be-idan ha-tikshoret ha-digitalit*, 106.

stable long-term employer-employee relations, as was the custom in Fordist times. Permanent in-house employees are being replaced with flexible temporary workers employed part-time, for a limited time, or on an "as needed" basis. The possibilities for a linear career are decreasing. Network workers must manage their own careers, with all the financial and social risks involved in it. The individualization of career management requires extensive self-promotion and advertisement of one's skills, abilities and qualifications. This has given rise to social networks like LinkedIn, which have become the "market square" where people (workers) can advertise, display, and offer their work skills for sale. From the aspect of personal privacy, it means displaying one's private sphere out in the public sphere.[336]

The workplace (whether home or office) has become a place of "research and development," "intellectual labor," "creative work," "symbolic analysis," and many other terms coined to describe network work. The detachment of work from time and space allows greater flexibility in choosing one's employers or customers, ignoring national borders, geographical distance and differences in time zones.[337] Digerati can bring more of themselves to work (their individual talents, creativity and initiative) and enjoy their work because they are involved in creative labor rather than monotonous production labor. Therefore, for an individual, the advent of new forms of work and production means more

---

[336] Castells, *The Rise of the Network Society*, 254.

[337] Thomas L. Friedman, *The World Is Flat: A Brief History of the Twenty-first Century* (New York: Farrar, Straus and Giroux, 2005): describes an outsource call center in Bangalore, India, which provides worldwide support services in multiple languages for several large Western companies.

opportunities to be involved and to use their individual talents, skills, passions and interests in the production process.[338]

The result is an invasion of the public sphere into the inner world (Popper's World 2) and vice versa – a display of the inner world out in the public sphere. The elements of personal information (including private information) of digerati that flow through the network are as follows:

Information category	Information elements	Popper classification
Body-related information	Physical traits and abilities.[339]	World 1
Mind-related information	Information about personality traits such as talents, initiatives, creative skills, education and expertise.	World 2
Information about actions	Actions and lifestyle habits characteristic of a specific digerati, such as work experience.	Worlds 2 and 3
Property information	The means of communication one possesses.	World 1
Information about relationships with external entities	What other entities in the production network one interacts with.	World 3
Information about the values that guide decisions and actions (autonomy information)	Values and norms which affect what one will or will not do.	Worlds 2 and 3
Identity-related information	Identifying information one is required to provide to become a node in the work network: passport identity and occupational identity.	World 1
Anonymity	Impossible for a node in the work network.	Worlds 2 and 3

---

[338] Fisher, *Capitalism be-idan ha-tikshoret ha-digitalit*, 180.

[339] E.g., information about physical disabilities that can influence one's ability to perform certain tasks.

The Network Market and Network Economy

Network production and network work are two building blocks of the network economy. The third one is the network market. Together, they create a new kind of economy – the informational economy.[340] The network market breaks the equilibrium between governments, corporations, and capitalists, which is characteristic of the Fordist social structure.[341] The Fordist economy is being replaced by a neo-liberal economy. The beginning of this process can be seen in Margaret Thatcher's policy of shifting the emphasis from state responsibility to individual responsibility and giving priority to initiative, incentives, and wealth generation.[342] According to Roland Kley, the new economy is essentially communication-based, with the Internet being its dominant technology, which allows every individual to become a node in the economic network and to communicate with other nodes in order to perform its financial operations.[343] Every individual node is "dumb," i.e., unable to see and comprehend the network in its entirety. At any given time, every person has only partial information about the market. Here lies the main difference between the digital-economic discourse and neo-liberal thought: while the neo-liberal philosophy frames these arguments under abstract theoretical conceptions (free competition, complete information about the market),[344] in the digital discourse, they are

---

[340] Castells, *The Rise of the Network Society*, 163.

[341] Jessop, "Fordism and Post-Fordism", 1-41.

[342] Daniel Yergin and Joseph Stanislaw, *The Commanding Heights: The Battle Between Government and the Marketplace That is Remaking the Modern World* (New York: Simon & Schuster, 1998), 139-140.

[343] Roland Kley, *Hayek's Social and Political Thought* (New York: Oxford University Press, 1994), 256-259, 262-263.

[344] Fisher, *Capitalism be-idan ha-tikshoret ha-digitalit*, 81.

translated into "material" components of the ICTs, such as microchips, network architecture, and software architecture such as big data. The new economy uses network technology to remove the obstacles (norms and regulations) between the individual and the market. In the digital discourse, the network market is described as chaotic but self-regulating and working best when left unsupervised, especially politically.

According to Chris Anderson, "Online markets are nothing if not highly efficient measures of the wisdom of crowds."[345] Network technology is great at harnessing and monetizing the non-creative power of the masses, and this means that **our personal information has become the essential fuel that powers the network economy**. The process of using personal information to fuel the network market has two stages, between which there is continuous feedback: the first stage is obtaining personal information, and the second stage is using it to power the industry. Shoshana Zuboff explains that the initial purpose of collecting personal information was to improve the quality of services,[346] but with time, Internet companies – first Google, then others such as Facebook and Netflix – began to realize that the "behavioral surplus" was an essential economic resource for them, because the aggregation and analysis of user information helped them accurately predict consumer wishes and preferences, allowing targeted advertising, and eventually also niche production and niche marketing ("long tail" production). That was the very resource that helped Internet companies thrive and survive the dot-com crash of the early 2000s. This, according to Zuboff, was the beginning of surveillance capitalism: "Google discovered that we are less valuable than others' bets on our future

---

[345] Anderson, *The Long Tail*, 223.
[346] Zuboff, *The Age of Surveillance Capitalism*, 70.

behavior."[347] This discovery has transformed the network market and network economy.

The mechanism of surveillance capitalism works as follows:[348]

(1) The Internet companies' nonmarket interactions with users are translated into surplus raw material; in other words, personal information we provide is used as a resource from which Internet companies manufacture the "prediction products" they sell to their real customers – advertisers This is a process in which each individual's online activities are used to create behavioral patterns, which are then analyzed to predict a person's other preferences based on their known preferences, and eventually to make production and marketing forecasts for marketed goods. In Zuboff's words, "Industrial capitalism transformed nature's raw materials into commodities, and surveillance capitalism lays its claims to the stuff of human nature for a new commodity invention." We are put under constant surveillance without our consent, and all that is left for us is to try and find some ways to hide.

(2) The means of production are AI technologies based on processing big data, which helps them constantly improve their prediction abilities. While in the industrial systems of the Fordist era there was a tension between quantity and quality, deep learning AI technologies improve more the more data they process, which means quantity is a condition for quality rather than an obstacle. Your Amazon click trails are used to create better recommendations for the next visitor. Your Google searches and the pages you visit provide feedback that helps improve the search algorithm's accuracy. The ads you click on not only increase Google's revenue but also tell it how much it can charge the

---

[347] Ibid., 93.
[348] Ibid., 93-97.

next advertiser. Internet companies have learned to harness information that had been there all the time, invisible and unexploited, and turn it into a bottom line.

(3) The products of surveillance capitalism are consumption predictions made by analyzing the behavioral surplus using deep learning technologies to forecast what we will feel, think, and do. Internet companies are telling the truth when they say they do not sell our data; what they do sell, however, is marketing predictions they fabricate from that data. Their claims of "privacy purity" are intended to hide their use of our personal information as the raw material for their industry. The more accurate the prediction, the more customers follow it, making it a self-fulfilling prophecy. This process supports long-tail production and makes traditional advertising, which in Fordist times served as a bridge between what is being produced and what will be purchased, redundant in many ways.

(4) The marketplace of prediction products is not limited to the market of physical goods; it includes the use of our predicted future behavior for political purposes. A good example of this is the Facebook-Cambridge Analytica data scandal.[349]

Zuboff describes surveillance capitalism as:

- A new economic order that claims human experience as free raw material for hidden commercial practices of extraction, prediction, and sales;

------

[349] As a result of which Facebook founder and owner Mark Zuckerberg was called to testify before the US Congress to provide explanations and present a plan of action for improving Facebook users' privacy protection.

- An expropriation of critical human rights and an overthrow of the people's sovereignty.[350]

If this is true, not only is personal information a resource provided (as of now) without a fee, it is used to create a new capitalist order that threatens the Fordist way of life in general and its conceptions of personal privacy in particular. Our every interaction with this mechanism makes it stronger and helps Internet companies increase their profit by commodifying our personal information. According to Zuboff, "This cycle will be broken only when we acknowledge as citizens, as societies, and indeed as a civilization that surveillance capitalists know too much to qualify for freedom."[351]

Since every individual is a node in the network, the network market is made up of many nodes, each of which has only partial knowledge about it. The knowledge that drives the network market toward equilibrium is the knowledge obtained by AI tools, particularly deep learning, from our personal information. This is a completely different perception of market behavior and balancing mechanisms. The free-market economy is based on the assumption that every player in the market has complete information about the conditions of the market, which is balanced by "the invisible hand" by aligning supply and demand around agreed prices. Nowadays, says Bruce Schneier, "we all have a customer score [...], and it's focused on what you buy, based on things like purchasing data from retailers, personal financial information, survey data, warranty card registrations, social media interactions, loyalty card data, public records, website interactions, charity donor lists, online and offline subscriptions, and health and fitness information. All

---

[350] Zuboff, *The Age of Surveillance Capitalism*, "The Definition", 8.
[351] Ibid., 499.

of this is used to determine what ads and offers you see when you browse the Internet."[352] He further adds: "Corporate surveillance and government surveillance aren't separate. They're intertwined; the two support each other. It's a public-private surveillance partnership that spans the world."[353] According to Zuboff, the power of surveillance capitalism is founded on six basic declarations, each of which builds on the one before it:[354]

1. Claiming human experience as raw material free for the taking, which allows its use while ignoring considerations of individuals' rights, interests, awareness, or comprehension.
2. This claim gives the right to translate individual people's experiences into behavioral data.
3. The right to own the behavioral data derived from human experience – based on 1 and 2.
4. The right of ownership confers the right to know what the data disclose.
5. The rights claimed in 3 and 4 confer the right to decide when and how to use the obtained knowledge.
6. The previously claimed rights create the conditions for the claim to preserve those rights.

Each of these six declarations alone – and all of them combined – undermines the value of human beings as autonomous individuals with clearly defined rights and obligations that were shaped in the modern era. According to Zuboff, these are the exact things Internet companies do:

---

[352] Schneier, *Data and Goliath*, 110.
[353] Ibid., 78.
[354] Zuboff, *The Age of Surveillance Capitalism*, 178.

they use our life experience and our ignorance of the knowledge they have about us to further increase that knowledge (i.e., to reduce our personal privacy) and to maximize their profits.[355]

Through these six claims for extensive unsupervised use of our personal information, surveillance capitalism undermines personal privacy and the privacy of personal privacy information, challenging the paradigm of personal privacy described in Chapter 7.

The following table describes the personal information (including private information) of market nodes that flows through the network:[356]

---

[355] Ibid., 498.
[356] The ways personal information is collected, aggregated, analyzed and disseminated online are explained in further detail in other sections.

Information category	Information elements
Body-related information	Medical information and information about physical properties are based on the analysis of our purchases of medications, health products, etc.
Mind-related information	Information about personality traits such as skills, talents, initiatives, education, creative ideas and hobbies.
Information about actions	Actions and lifestyle habits characteristic of a specific individual.
Property information	What one owns – based on one's online activity.
Information about relationships with external entities	For example, information about one's familial relationships based on blogs, Facebook contact lists, WhatsApp groups and greetings one receives.
Information about the values that guide decisions and actions (autonomy information)	Political opinions, values and norms that affect one's surfing patterns.
Identity-related information	Passport identity information required and used for identification and sometimes for the app itself.
Anonymity	Impossible for a node in the network market, mainly because physical goods (but not just them) purchased online are meant to be shipped to a specific physical person.

From this overview of the network market and surveillance capitalism, we can learn that personal information (including private personal information) has become the raw material of the network market economy and is being used by the players in this market for profit or to achieve social and political goals. The use and commodification of our personal information takes it out of our control and puts it under the control of the players in the network market. By definition, the network market resides in the public sphere (except in certain cases where the information exchange is between confidants), which, driven by the market's economic and political interests, penetrates the private sphere until the distinction between the two is blurred. As a result, increasing amounts of personal privacy information and deep personal privacy information (Popper's World 2) become exposed to the public.

The Economics of Privacy (Personal Privacy as an Economic Resource)

Information economy has developed in the 20th century as part of the rising Third Wave.[357] One of its subsets is the economics of privacy, a field that focuses on the economic changes resulting from the protection or disclosure of private personal information.[358] Disclosure of personal information cannot be categorized as entirely good or entirely bad for the individual. In the next sections, I will describe the pros and cons of information disclosure and/or non-disclosure for the individual using economic models that regard private information as an economic resource.[359]

---

[357] Toffler, *The Third Wave*.

[358] Alessandro Acquisti, Curtis Taylor and Liad Wagman, "The Economics of Privacy", *Journal of Economic Literature* 54 (2) (June 2016), 442-492.

[359] Including descriptions of specific economic models that lie at the basis of network economy, which uses private information as a valuable economic resource.

The Economic Value of Personal Information

The widespread use of web applications allows the collection of personal information that before the Internet was inaccessible and remained in the private sphere. For example, the transition from linear TV watching – watching a film or a show on a specific channel at the specific time when it is broadcast – to video-on-demand services like Netflix allows to collect information about our viewing habits: when we watch TV, the genres we prefer, and how many times we rewatch the same movie or episode. This, combined with other data, allows the creation of a viewer profile accurate enough to trade with other Internet companies that can use it for targeted advertising. The advent of the Internet led to the formation of Internet conglomerates – such as Facebook, Google and Amazon – that provide social apps and e-commerce services, which means they are sitting on a gold mine of personal information. That information is stored, processed, analyzed and shared by and between them at an unprecedented scope to maximize their profits. Disclosure or non-disclosure of your personal information has both positive and negative financial implications – not only for you but also for the entities you are (or are not) dealing with.[360] For instance, if a bank or a credit company has some information about your finances, it can help you get a loan at terms that are good for you or for the bank – or prevent you from getting one. Similarly, disclosure and non-disclosure of personal information have benefits and drawbacks for society, such as uncovering terrorist activity, tax evasion, or disclosure of individuals' COVID-19 vaccination status. Alessandro Acquisti sums it up thus: "Personal data

---

[360] See Appendices A and B for the game theory models I offer as a way to balance the pros and cons of disclosure and/or non-disclosure of personal information.

is continuously bought, sold, and traded among firms (from credit-reporting agencies to advertising companies to so-called "infomediaries," which buy, sell, and trade personal data), but consumers themselves do not have access to those markets: they cannot yet efficiently buy back their data, or offer their data for sale, although the concept of personal-information markets for consumers, or individuals' markets for privacy, has been around since the mid-1990s".[361]

This raises a number of questions:

- Can the benefits of personal information disclosure for the information subject and the other parties to the information exchange be balanced against its drawbacks?
- How do Internet companies profit from our personal information?
- What would be the right balance between the economic value of personal information that is being aggregated in big data pools and traded in conditions of a free market that regards privacy as an economic commodity and legislation and social norms that regard privacy as a fundamental human right?

To answer these questions, let us examine the economic value of personal information, its implications for personal privacy, and consumers' understanding of the benefits and costs of their decisions regarding the protection or disclosure of their personal information. As mentioned earlier, maintaining privacy may be more or less beneficial for you, depending on your circumstances and various other conditions. The

---

[361] Acquisti, Taylor and Wagman, "The Economics of Privacy", 447.

extent of its benefit for you may correspond with the extent of its benefit for society or contradict it. However, in most cases, people don't have enough information regarding the benefits or costs of disclosing or maintaining the confidentiality of their personal information when they make transactions online. Furthermore, any benefits they achieve in return for their personal information are by nature temporary and not guaranteed. Personal privacy (and, consequently, personal privacy information) has several basic components,[362] some of which cannot be evaluated in terms of financial or material consideration;[363] therefore, in this section, I will only discuss those components of privacy that can be translated into material consideration. Acquisti describes three waves of research in the economics of privacy:[364]

The first wave focused mainly on the question of whether withholding personal information negatively impacts free competition in a developed market. Scholars argued that it prevents decision-makers from accessing information that could help them make informed decisions, as in the case of providing a loan without full knowledge of the loaner's financial picture or hiring a person for a job without having accurate information about their abilities and qualifications.

The second wave emerged with the advent of the Internet, and the economic implications of the secondary uses of personal information resulting from the sharing of personal information online were studied. One of the expressed concerns was that the amount of shared personal information would be too small to satisfy consumer needs: people may

---

[362] This book defines ten.

[363] For example, can one quantify or put a price on the damage caused by shaming as a result of exposing a person's sexual preferences without their consent?

[364] Acquisti, Taylor and Wagman, "The Economics of Privacy", 450-467.

want their preferences regarding certain commodities to be known so they could receive more relevant offers. On the other hand, it was noted that too much information about a consumer's weaknesses may lead to abuse of those weaknesses: e.g., a person may be offered more expensive credit deals based on the knowledge that they are temporarily short of cash. Another concern expressed by the second-wave researchers was that when people provide their personal information (such as their bank account number) in a credit card transaction, they have no knowledge or control over how and by whom that information will later be used. One of the views that took hold in the second wave of research was that privacy protection was not merely a regulatory issue but an economic one as well, resulting in visions of "information markets" where individuals own their personal data and can transfer the rights to that data to others in exchange for some type of compensation. It was acknowledged that such "markets" would require appropriate legislation and regulation, which means the second-wave scholars considered personal information trading and regulation of personal information use to be complementary rather than contradictory. Under this conception, what remained to be discussed was the appropriate equilibrium between a free market of information and privacy regulations that would restrict and moderate it. Such an equilibrium would be time- and culture-dependent.[365]

The third wave was the result of the proliferation of the Internet, which included targeted marketing and advertisement based on

---

[365] Ibid., 453-454.

The article gives an example of an unregulated market controlled by a monopoly, which reaches equilibrium when the consumers sell their personal information for a price of 0, with the monopoly being the only one that profits from that.

behavioral surplus analysis. Such use of personal information for commercial and political purposes is very widespread on the Internet. For the first time, our personal information is closely connected with our commercial activity. Personal information has several economic aspects.[366]

Different Models of The Economy of Personal Privacy

Several economic models consider merchants' ability to practice price discrimination against consumers based on personal information regarding their preferences. Those models are asking: Is there any regulatory protection of personal information? Are consumers aware of the fact that by using ICTs, they are providing merchants with personal information about themselves, such as their preferred products, shopping places and shopping times, which can help the merchant maximize its profits? **At the basis of those economic models lies the assumption that the first component of personal information privacy to be violated is anonymity.[367] The consumer is not anonymous for the merchant, and once anonymity is broken, it becomes easier to violate and exploit their preferences (and weaknesses) in additional components of privacy, such as information about actions (previous purchases) or body-related information (gender, specific health**

---

[366] I will try to explain the economic aspects of each of the basic components of privacy.

[367] As described in Chapter 1, the main technological tools used to violate anonymity are cookies and fusion of information from different big data pools.

conditions).[368] According to Curtis R. Taylor, in a setting where merchants have access to consumers' personal information, they will use it to price discriminate.[369] The usefulness of regulatory protection of information privacy depends on consumers' level of sophistication: merchants may take advantage of naïve consumers who are unaware of the uses made of their personal information; therefore, such consumers need regulatory intervention to protect their personal information. However, sophisticated consumers who are aware of the ways merchants use personal information can adapt their purchasing behavior to make it in the company's interest to protect their data, thus making regulation unnecessary.[370] For example, you can skip a purchase today to avoid being identified as a past customer tomorrow – and thus have access to lower prices targeted at new consumers.[371] Different studies show how disclosure or protection of personal information may affect the competitive structure of a market. According to a study by Campbell, Goldfarb and Tucker, if the regulation is based only on one-time consumer consent to the use of their data, it may generate market power and monopoly because consumers would be more willing to provide such consent to large companies than they would to small ones.[372] Thus, regulation focused mainly on imposing consumer consent in advance

---

[368] For example, offering a woman pregnancy and baby products knowing she is pregnant.

[369] Curtis R. Taylor, "Consumer Privacy and the Market for Customer Information", *The RAND Journal of Economics* 35 (4) (February 2004), 631-650.

[370] For an example of this model, see Appendix B.

[371] Acquisti, Taylor and Wagman, "The Economics of Privacy", 455.

[372] James Campbell, Avi Goldfarb and Catherine Tucker, "Privacy Regulation and Market Structure", *Journal of Economics and Management Strategy* 24 (1) 2015, 47-73.

favors large Internet companies over small ones and may be anti-competitive. A study by Kim, Wagman and Wickelgren examined the impact of horizontal mergers on market competition for companies that, prior to the merger, employed price discrimination based on individual preferences[373] and found that such "individualized pricing" practices helped decrease competition even further after the merger. Another study by Armstrong, Vickers and Zhou found that in a duopoly model[374] that only offers discounts to new customers (online or offline), customers do not return.[375] This finding explains Amazon's 2001 commitment not to employ such policies.[376] The conclusion from these studies is that large companies with great market power often benefit from committing to protect their consumers' personal information.

Information Mediation

This section will focus on the role of large Internet companies such as Google, Facebook, Amazon and others, which collect our personal

---

[373] Jin-Hyuk Kim, Liad Wagman and Abraham L. Wickelgren, "The Impact of Access to Consumer Data on the Competitive Effect of Horizontal Mergers and exclusive dealing", *Journal of Economics & Management Strategy,* 28 (3) (Fall 2019), 373-391.

[374] A duopoly is a type of oligopoly where two vendors have dominant or exclusive control over a market. Two companies that control the majority of the market are considered a duopoly de facto, even if there are additional competitors.

[375] Mark Armstrong, John Vickers and Jidong Zhou, "Consumer Protection and the Incentive to Become Informed", *Journal of the European Economic Association* 7 (2-3) 2009, 399-410.

[376] Simon P. Anderson and Andre de Palma, "Competition for Attention in the Information (Overload) Age", *The RAND Journal of Economics* 43 (1) (Spring 2012), 1-25.

information and mediate (trade) it to merchants and advertisers – thus, in fact, creating surveillance capitalism, and present some detailed economic models of this market. According to De Cornière and Nijs, the online advertising industry has made it possible to customize advertising by using private personal information[377] and to segment the market down practically to the level of individual people, with an almost complete match between the consumer's preferences and the merchant's supply. This is often accompanied by adjustments of prices (price discrimination). A study by Board and Lu describes how this affects the competitive structure of the market:[378] on the one hand, consumers search between different merchants for the deal or product that would be best for them; on the other hand, merchants collect and analyze that search information. There are two models of personal information use: in the first model, the merchants know the consumers' preferences, and the equilibrium point results from a "monopoly on information," which merchants use to offer products that are more profitable for them but not necessarily best suited for the consumer. In the second model, the consumers remain anonymous, and their preferences remain private; the equilibrium point of this model is when the merchants display complete information about their products. These and other studies[379] of the economic impact of information mediation show that Internet companies that mediate personal information normally do not act in the consumers'

---

[377] Alexandre de Cornière and Romain de Nijs, "Online advertising and privacy", *The RAND Journal of Economics*, 47 (1) (Spring 2016), 48-72.
[378] Simon Board and Jay Lu, "Competitive Information Disclosure in Search Market", *Journal of Political Economy* 126 (5) (August 15, 2015), 1965-2010.
[379] Acquisti, Taylor and Wagman, "The Economics of Privacy", 459.

best interest. For a more detailed description of economic models that demonstrate this, see Appendices A and B.

Marketing Techniques

Some studies extend their analysis to the costs of unauthorized intrusions into consumers' private space – including spamming, junk mail and pop-up ads.[380] Those studies have found that price discrimination pushes companies to a "price war," which results in less accurate targeted advertising, differentiated investment in client hunting, and eventually prioritizing customer loyalty over price discrimination.[381] This shows that merchants are better off using direct marketing than employing intrusive techniques. Consumers know how to protect themselves against those techniques (mainly by ignoring the messages, directing them to the Spam folder or deleting them), which makes them fairly ineffective. According to the studies, merchants prefer to use different advertisements for different categories of consumers, selecting the information they deliver and the product traits they emphasize based on the preferences of each particular segment. In Appendix B, I describe a marketing and spamming model.

Privacy, advertising and e-commerce: Targeted advertising based on consumer profiling is one the most widespread practices

---

[380] "Spam" or "spamming" is a general term for nuisance contents communicated to a person against their will, and includes electronic communications as well as regular mail, automatic phone calls and fax messages. "Junk mail" or "junk messages" refers to spam contents communicated electronically – usually by email, but also by text messages or instant messaging apps such as WhatsApp.
[381] Increasing or reducing prices according to the merchant's needs.

employed by Internet companies today. This online advertising practice is the basis of surveillance capitalism and the source of Internet companies' wealth.[382] Consumers are profiled using an arsenal of surveillance and monitoring tools that monitor them continuously both in the physical and virtual spheres.[383] Even though targeted advertising has the theoretical benefit of maximizing the match between consumer preferences and the selection of manufactured products, thus preventing waste of resources, in reality, it is often used to take advantage of people's weaknesses. Furthermore, individuals are usually unaware that they are being monitored, and even if they are aware of it, they cannot avoid it.[384] One example of a service that collects personal information with nothing the consumer can do to prevent it is mobile payment systems and shopping apps. So far, none of the regulatory attempts to supervise and minimize the privacy violations involved in targeted advertising have been sufficiently effective.[385] The financial impact of

---

[382] In the first quarter of 2021, Google beat analyst expectations with a revenue of \$55.31 billion, \$44.68 billion of which (an increase of 32.3% compared to the previous year) came from advertising products: \$31.88 billion from "Google Search & other", \$6 billion from "YouTube ads", and \$6.8 billion from "Google Network".
https://www.sec.gov/Archives/edgar/data/1652044/000165204421000018/googexhibit991q12021.htm
https://www.cnbc.com/2021/04/27/alphabet-goog-earnings-q1-2021.html
(Accessed May 3, 2024)

[383] The relevant ICTs are described in Chapter 1.

[384] Chris Jay Hoofnagle, Jennifer M. Urban and Su Li, "Mobile Payments: Consumer Benefit & New Privacy Concerns", *SSRN Electronic Journal*, April 24, 2012. https://papers.ssrn.com/sol3/papers.cfm?abstract_id=2045580
(Accessed 4/5/2021)

[385] E.g., the GDPR.

restrictive regulation on the producers, advertisers and consumers is ambiguous. However, consumers tend to prefer unobtrusive targeted ads to obtrusive ones, regardless of the actual extent of personal privacy violation.

Privacy and price discrimination: Since there is no clear evidence of real-time online price discrimination, I have nothing to add on the subject besides what has already been said in the previous sections.[386]

Other forms of discrimination: As discussed in the previous sections, disclosure or non-disclosure of personal information can have both a positive and a negative side. Anonymous exams help make grading fairer; maintaining maximum anonymity when applying for a job can help prevent discrimination based on gender, faith or ethnicity; on the other hand, hiding your criminal record can harm your potential employer. Dating apps are another example of the positive and negative sides of sharing or hiding personal information. Thus, the advantages and disadvantages of discrimination based on personal information are context-dependent. Some prominent examples of economic systems based on sharing personal information are platforms like Airbnb and Uber Taxi.

Privacy and health economics: There are several difficulties associated with the protection of medical data privacy:[387]

Patient data collected by healthcare providers, hospitals, healthcare organizations and insurance companies are being digitized and stored in big data pools to facilitate treatment, follow-up and research. However, this also facilitates a potential breach of medical

---

[386] Acquisti, Taylor and Wagman, "The Economics of Privacy", 466-467.
[387] Ibid., 469-471.

confidentiality, which is part of personal privacy. Information that can help link medical data with a specific person, which had only been available to confidants, is now publicly accessible. Currently employed technological tools and practices such as data deidentification are insufficient to prevent it.[388] Biological data pools, which contain individual persons' biological characteristics, allow the identification of specific people. Genetic information that has clinical significance may also be used to profile a person.

The economic significance of medical information raises the question of who owns the medical information – the information subject (the patient) or the entity that creates and controls the database (usually a medical institution). Personal medical information that is supposed to be confidential is traded[389] and used for profiling (e.g., by insurance companies). This clearly indicates a violation of medical information confidentiality.

Privacy and credit markets: Credit card transactions are subject to extensive regulation designed to protect the confidentiality of the personal information required for making such transactions.[390] But even so, consumers have little or no power to prevent the sharing of their information with affiliated and non-affiliated companies (e.g., as part of a loan approval). Sometimes, the consumer has to actively opt in to allow the use of their personal information; sometimes, they can only opt out retroactively.

The advent of ICTs gave rise to big data pools, which have become a separate ecosystem that allows storage, fusion, analysis and

---

[388] See discussion of *k*-anonymity in Appendix C.

[389] For example, the medical data of millions of Israelis were shared with Pfizer as part of the COVID-19 vaccination purchase agreement.

[390] Acquisti, Taylor and Wagman, "The Economics of Privacy", 471-473.

dissemination of personal information as part of economic models that treat it as a commodity. The market of personal information is controlled by a handful of Internet corporations such as Facebook, Google, Amazon and a few more. The consumers are provided with free products and services, and in return, their personal information is made into a commodity traded on the network market. Legislation does not keep pace with this phenomenon.[391] Internet companies justify it by claiming they are serving the common good and contributing to an open society in which both consumers and merchants have access to the right selection of options so that each can make the best decisions for themself.[392] This argument is used to justify targeted advertising, which supposedly bridges a consumer's preferences and the produced supply. The Discrepancy Between Consumers' Attitudes toward Information Privacy and Actual Behaviors.

Many empirical studies have found a puzzling dissonance between people's awareness and concern for their privacy and its violation by Internet companies, as well as their actual consumer behavior.[393] When required to choose between getting a free product or service from an Internet company in exchange for providing personal

---

[391] The GDPR is a step toward regulation of the information market.

[392] From Mark Zuckerberg's speech at Georgetown University: "I remember feeling that if more people had a voice to share their experiences, maybe things would have gone differently. Those early years shaped my belief that giving everyone a voice empowers the powerless and pushes society to be better over time". Appeared in: Tony Romm, "Zuckerberg: Standing for Voice and Free Expression", *The Washington Post*, 17 October, 2019 https://www.washingtonpost.com/technology/2019/10/17/zuckerberg-standing-voice-free-expression/ (Accessed 4/5/2021)

[393] Acquisti, Taylor and Wagman, "The Economics of Privacy", 476-478.

information and not getting the product or service, most people would choose to provide their personal information. According to a 2009 study, younger Americans are less likely to say no to targeted ads than older Americans, yet 55% of 18-24-year-olds say they do not want tailored advertising. 86% of the participants say they don't want tailored advertising if it is the result of following their behavior on other websites, and 90% reject it if it is the result of following what they do offline (e.g., by means of security cameras).[394] This apparent dichotomy raises several questions: How much does privacy really matter to people? Can the value of personal privacy protection be quantified financially, and how? There are several possible resolutions to the paradox. The first one is the theory voiced by Internet companies' officials: as formulated by Calvin Gottlieb, "Most people, when other interests are at stake, do not care enough about privacy to value it."[395] I personally believe such an underestimation of personal privacy compared to other interests is a misrepresentation of the choices people make. There is no real discrepancy between perceiving personal privacy as an essential need and seemingly waiving it, as described above. Instead, each particular choice is a result of weighing the pros and cons of maintaining personal privacy versus waiving it. Furthermore, the need to ensure the legal protection of privacy rather than leave it to people's own momentary

---

[394] Joseph Turow, Jennifer King, Chris Jay Hoofnagle, Amy Bleakley and Michael Hennessy, "Americans Reject Tailored Advertising and Three Activities That Enable It", *SSRN Electronic Journal*, 29 September 2009. https://papers.ssrn.com/sol3/papers.cfm?abstract_id=1478214 (Accessed 5/5/2021)

[395] Gottlieb, "Privacy: A Concept Whose Time Has Come and Gone", 156.

decisions led to the various laws and regulations intended to confront personal privacy violations online.[396]

The other possible resolution is that people – whether consciously or unconsciously – routinely engage in mental trade-offs or a so-called "privacy calculus." This is not a zero-sum game, i.e., receiving online services does not imply a complete loss of privacy; rather, these are constant cost-benefit calculations of the pros and cons of receiving a particular online service in exchange for giving up some of your privacy. However, one of the arguments I am making in this book is that any such "calculus" is initially wrong because of the ICTs' ability to put together all the different pieces of information to create a complete picture of our personal information, in extensive violation of our personal privacy as a whole. While you may think you are only giving up a few pieces of your personal privacy for each online service (or app) you use, behind-the-scenes ICT tools fuse all those separate pieces of information to create an extensive database so that eventually, Internet companies hold much more of your personal privacy information than you could ever imagine you allowed them to.[397] This is the information asymmetry between individuals and Internet companies.[398]

---

[396] E.g., the 1980 OECD Guidelines on the Protection of Privacy and Transborder Flows of Personal Data, the EU GDPR, Israel's Data Protection Law, and so on.

[397] See Chapter 10 for a discussion of the problem of consent and control.

[398] The Facebook-Cambridge Analytica data scandal is a great example of this asymmetry. See: Keith Collins and Gabriel J.X. Dance, "How Researchers Learned to Use Facebook 'Likes' to Sway Your Thinking", *New York Times*, March 20, 2018 https://www.nytimes.com/2018/03/20/technology/facebook-cambridge-behavior-model.html (Accessed May 06, 2024)

Different studies have attempted to put a price tag on privacy. One such study found that elements of users' browsing histories are being traded among Internet advertising companies for less than \$0.0005 per person.[399] According to another study, Americans are willing to pay an average of \$37 for protection against improper use of and trading in their private information.[400] These two studies reveal the enormous gap between the value people attribute to their personal information and the price paid for it by Internet advertisers.

Regulation versus Self-Regulation

Various empirical studies of privacy trade-offs have contributed to the debate over how to best protect privacy without compromising the beneficial effects of information sharing.[401] Much of this debate juxtaposes the relative benefits of regulation and self-regulation. Robert Gellman estimates that in 2001, \$18 billion were lost by companies in Internet retail sales due to buyers' privacy concerns, and appraises that consumers pay a much greater price when their privacy is not protected (including losses associated with identity theft, differential pricing, spam, and investments aimed at protecting data).[402] This side of the

---

[399] Lukasz Olejnik, Tran Minh-Dung and Claude Castelluccia, "Selling Off Privacy at Auction", Paper presented at ISOC Network and Distributed System Security Symposium, San Diego (February 2014), 23-26.

[400] Il-Horn Hann, Kai Lung Hui, Sang-Yong Tom Lee and Ivan P. L. Png, "Overcoming Online Information Privacy Concerns: An Information Processing Theory Approach", *Journal of Management Information Systems* 24 (2), 13-42.

[401] Acquisti, Taylor and Wagman, "The Economics of Privacy", 479-481.

[402] Robert Gellman, "Privacy, Consumers and Cost: How the Lack of

debate advocates regulatory solutions to privacy problems. On the other hand, some scholars argue that the economic benefit of targeted advertising for merchants and consumers alike outweighs the costs mentioned earlier.[403] This side of the debate advocates self-regulation, i.e., that each Internet company should adhere to its own privacy policy. Self-regulation relies on transparency in the companies' use of personal information and individuals' ability to accept or reject it without negative repercussions, i.e., leaving the control over individuals' personal information in the individuals' own hands.

Big Data and Privacy-Enhancing Technologies

The advent of big data pools has opened up new opportunities for using personal information as an economic resource with the aid of cutting-edge technologies such as data mining, deep learning and the Internet of Things,[404] and taking it from "just" targeted advertising to the next level – affecting our minds and reducing our autonomy.[405] This has led to the development of privacy-enhancing technologies and mathematical

---

Privacy Cost Consumers and Why Business Studies of Privacy Cost are Biased and Incomplete". https://epic.org/report/dmfprivacy.html (Accessed 9/5/2021)

[403] Paul H. Rubin and Thomas M. Lenard, *Privacy and the Commercial Use of Personal Information* (New York: Springer, 2001), 1.

[404] Acquisti, Taylor and Wagman, "The Economics of Privacy", 481-482.

[405] Jeff Orlowski's documentary *The Social Dilemma* premiered on Netflix in 2020, sparking a heated public debate. While the fact that we are being monitored by social media is old news, this film exposes social networks' use of AI algorithms designed to create addiction with a sole purpose of maximizing the social media companies' profit.

principles such as *k*-anonymity and differential privacy.[406] The question remains as to the individuals' gains and losses associated with maintaining privacy versus the social benefit of disclosing personal privacy information.[407]

The Network Human

In Chapters 4 and 5, I described how the Western conception of individualism was shaped starting from the Age of Enlightenment and through the Industrial Era and how the conception of personal privacy evolved from that. The information and communication technologies described in Chapter 1 are developing rapidly in accordance with Moore's law, and their use is spreading rapidly as well,[408] affecting every aspect of our lives. The age of ICTs has created a new culture, which, according to Chris Shilling, is creating new human experiences that necessarily change the way we perceive our bodies and our identities.[409]

According to Luciano Floridi, the rapid advent of ICTs and their impact on our lives force us to reconsider some of the most important components of our worldview:[410]

---

[406] Kobbi Nissim, "Differential Privacy: Why, How and Where to?", delivered at: Privacy in Challenging Times: The 8th Technion Summer School on Cyber and Computer Security (September 2020).

[407] For example, maintaining a patient's anonymity in a medical database may hamper the quality of their medical care.

[408] Ilya Levin and Dina Tsybulsky, "The Constructionist Learning Approach in the Digital Age", *Creative Education* 8 (2017), 2466.

[409] Chris Shilling, *The Body in Culture: Technology and Society* (London: SAGE Publications, 2004), 181-199.

[410] Luciano Floridi, "Introduction", in: *The Onlife Manifesto: Being Human in a Hyperconnected Era*, ed. Luciano Floridi (United Kingdom:

- Our self-conception (who we are);
- Our mutual interactions (how we socialize);
- Our conception of reality (our metaphysics); and
- Our interactions with and impact on reality (our agency).

Each one of these affects our conception of personal privacy and how we live it out in the world. Our self-conception determines the way we define our private sphere, i.e., the sphere where we are autonomous, free, independent etc. The nature and intensity of our mutual interactions define us as social beings bound by social norms, but they also define the boundaries between the private and public spheres and the information that does and does not flow through our social network. The terms used in this book define our social distance.[411] The challenges posed by ICTs are reshaping our conception of personal privacy due to at least four fundamental transformations that are taking place in our society in the digital age.[412] The transformations described by Charles Ess are as follows:[413]

A) Shifting from the primacy of entities over interactions to the primacy of interactions over entities

Before the age of ICTs, the emphasis was placed on the natures and characteristics of entities, while the interactions between them were

---

Springer, 2015), 1-6.

[411] See Chapter 2 for the metric for calculating social distance.

[412] Charles Ess, "At the Intersections Between Internet Studies and Philosophy: 'Who Am I Online?'", *Philosophy & Technology* 25 (2012), 275-284; Levin and Tybulsky, "The Constructionist Learning Approach...", 2463-2475.

[413] Ess, "At the Intersections Between...", 90. I chose to list the transformations in the order that I found most relevant to the purposes of this book, rather than in the original order used by Ess.

given secondary importance. However, in the age of ICTs, when every individual is a node in the network, their interactions with other nodes are becoming increasingly important. The other nodes may be individuals, commercial companies, government agencies, and organizations of different types. The increasing significance of interactions affects our self-conception and our agency. According to the actor-network theory, everything in the social and natural worlds is defined by its constantly shifting networks of relationships, and nothing exists outside those relationships.[414] Therefore, if the network human defines themself by their network, if they are at the same time a node in the network and a network agent, they will also define their personal privacy in terms of the network – i.e., which personal and private personal information will flow through the network. According to Pierre Lévy, the network activities of all the different individuals put together make the network similar to a living brain that has "collective intelligence." Alex Pentland goes even further, arguing we acquire most of our personal beliefs and social habits through social learning, by observing peers' positions, actions and consequences, rather than through our own logic. Social learning and maintenance of social contracts allow groups of people to coordinate their activities more efficiently, but it also means our rationality is, to a large extent, determined by the social fabric that surrounds us.[415] That is why, according to Pentland, privacy is so important: social influence is a powerful force that can lead people to great deeds – or terrible ones. It

---

[414] Fabian Muniesa, "Actor-Network Theory", in: *The International Encyclopedia of Social and Behavioral Sciences*, 2nd ed, ed. James D. Wright (Oxford: Elsevier, 2015), Vol. 1, 80-84.

[415] Alex Pentland, "The death of individuality: What really governs your actions?", *New Scientist*, 222 no. 5 (April 2014), 30-31.

can modify our behavior, and without privacy, its power is virtually unlimited. Eran Fisher supports these claims, saying the digital discourse challenges the modern conception of human beings as unique, distinct and autonomous individuals.[416] To wrap it up, the network culture, which has shaped the network human, is challenging the liberal Western conception of individuality and, consequently, the paradigm of personal privacy that stems from it. The following table summarizes the impact of the shift from the primacy of entities to the primacy of interactions on the basic components of our worldview:

Main changes in the age of ICTs Basic worldview components	The primacy of interactions between network nodes over the nodes themselves.	Resulting impact on various components of personal privacy.
Our self-conception (who we are)	A decrease in the importance we attribute to our selfhood and an increase in the importance we attribute to our place in the network.	Autonomy, anonymity, sense of subjectivity.
Our mutual interactions (how we socialize)	Increased scope and range of interactions (small world)	Reduced control over personal information. Personal information flows through the network with virtually no supervision.
Our conception of reality (our metaphysics)	The conceptions of space and time change as network location becomes more important than physical location in space and time.	Blurred distinctions between private and public spheres, network interactions free from space and time constraints.
Our interactions with and impact on reality (our agency)	Unprecedented ability to act anywhere, anytime, and to directly interact with any other node in the network.	Unmediated political and economic relations (direct democracy, prosumption) require the free flow of private personal information through the network.

---

⁴¹⁶ Fisher, *Capitalism be-idan ha-tikshoret ha-digitalit*, 185.

B) Blurring of the distinctions between human, machine and nature. Traditionally, people were able to distinguish between human and non-human beings (living or artificial). The age of ICTs brings a new approach, according to which machines are becoming indistinguishable from humans.[417] The line between human and machine is increasingly blurring. The machines of the age of ICTs are more than just extensions of our physical capacities; they are connected to us in a way that more and more resembles human-machine convergence. This affects our self-conception, our mutual interactions, our conception of reality and our interactions with and impact on reality. Thus, the blurring of the distinctions between human and machine radically challenges the humanist liberal conception of the individual. The new conception of the individual as an information entity[418] is one of the foundations of the blurring between humans and ICT and one of the things that enables their convergence: in Chapter 8, I have shown how personal privacy can be converted to information, including concepts related to our body, our mind and our social relations. One element of the blurring of the distinctions between humans and ICT is human-computer symbiosis, which J. C. R. Licklider defined as working together in intimate association in order to make decisions and control complex situations.[419]

---

[417] Floridi, *The Onlife Manifesto: Being Human...,* 1-6.
Based on the Turing test – a test of a machine's ability to exhibit intelligent behavior equivalent to – or indistinguishable from – that of a human, proposed by Alan Turing in 1950. See: A. M. Turing, "Computing Machinery and Intelligence", *Mind* 49 (1950), 433-460.
[418] A belief that the various traits and characteristics of a person, such as those that describe one's body, mental traits, social relations and identity, can be translated into elements of information, as I have shown in Chapter 8.
[419] J. C. R. Licklider, "Man-Computer Symbiosis", *IRE Transactions on*

In the age of ICTs, we can see numerous examples of physical human-computer symbiosis – such as nearly autonomous medical systems that can control a patient's blood sugar or heart rate or imaging devices that can monitor brain waves and discover conditions like cancer – as well as virtual human-machine symbiosis systems used to solve more abstract problems, such as AI-assisted stock trading or Waze (a navigation app where the human user sets the trip destination and starting time, and lets the machine choose the best route, which it may change in real-time based on the conditions of the road). All these are examples of human-computer symbiosis with a blurred distinction between the two. Another example of symbiosis is virtual assistants – such as Apple Siri or Amazon Alexa – that can learn what we like, what we want, and what we need. Can even learn our personality and proactively offer services or perform chores without being asked to. All those abilities are based on our private personal information that is being stored, processed and disseminated through the Internet.

The latest development in the field of human-machine convergence is apps (bots) that can imitate a conversation with a person who has passed away based on that person's private personal information stored over the years in big data pools.[420] Of course, such apps can also be used to imitate any living person as long as the app provider has enough information about that person.[421] There is a process of mutual learning between humans and machines, which may result in a blend and

---

Human Factors in Electronics, HFE-1 (1960), 1-21.

[420] Karine Alada, "Zal Digital" [The digital dead] (in Hebrew), *Haaretz Gallery*, 29 January 2021, 8.

[421] Ibid.; "Microsoft has been granted a patent for a chatbot that can take on the personality of real people" https://www.businessinsider.com/microsoft-patents-a-chatbot-to-imitate-real-people-2021-1

creation of a new type of being – the cyborg: "a cyber-organism, a hybrid of organic and machine body parts. It is a social being and an imaginary being".[422] According to Donna J. Haraway, we were used to seeing machines as distinguishable from humans because they lacked basic human characteristics such as autonomous movement, self-designing and the ability to achieve dreams.[423] However, these and many other distinctions are becoming ambiguous as we are faced with cyborg entities. ICTs are granting machines more and more human capacities, including autonomy, spontaneous movement and self-designing abilities.[424] This has major impact on personal privacy because the machines, which reside in the public sphere (databases stored in various cloud applications), are engaged in extensive communication and exchange of information, much of which is private personal information being transferred from the private sphere to public cloud services.

Human-computer symbiosis, now seen in a growing number of shared tasks and apps, is facilitating the back-and-forth switch of roles between humans and machines to the point where the distinctions between them begin to blur, and cyborg beings are created. It is a two-sided process in which we create technologies, including information and communication technologies, to be our extensions, but by using them, we are ourselves transformed.[425] Symbiotic systems are integrally built into the Internet network as a tool for transferring and processing enormous amounts of information while creating and using big data

---

[422] Alada, "Zal Digital".

[423] Donna J. Haraway, *A Cyborg Manifesto* (University of Minnesota Press, 2016), 11.

[424] E.g., neural networks' self-tuning control.

[425] Ess, "At the Intersections Between Internet Studies...", 94.

pools to allow rapid real-time processing of personal information.[426] In fact, it seems that human-computer symbiotic systems cannot exist without fast global communication networks. **Thus, the ever-growing human-machine symbiosis requires the transfer of personal information from the private sphere to the public sphere.** According to Eran Fisher, the blurring of the distinctions between human beings and network technologies means that our intelligence, creativity, selfhood, identity and experience are no longer limited to our physical person but spread out into the network,[427] exposing general and deep personal privacy information to the public.

The following table summarizes the impact of this transformation on the basic components of our worldview:

---

[426] "Real time" is the time window during which one can receive the information or response that one needs to solve a particular problem.

[427] Fisher, *Capitalism be-idan ha-tikshoret ha-digitalit*, 225.

Main changes in the age of ICTs Basic worldview components	Blurring of the distinctions between human, machine and nature	Resulting impact on various components of personal privacy
Our self-conception (who we are)	Our selfhood is no longer contained within our physical person alone; parts of it reside in the network, creating a cyborg personality.	Elements of our deep privacy (World 2) spread out into the public sphere. Personal privacy is dispersed all over the network.
Our mutual interactions (how we socialize)	Intimate interactions with machines, to the point of convergence and creation of cyborgs	Cyborg avatars that interact with the world in our name, which requires that they possess our private personal information even though they operate in the public sphere
Our conception of reality (our metaphysics)	Reality is becoming populated by a new type of beings – cyborgs, blends between humans and machines.	Blurring of the distinctions between human and machine also blurs the distinctions between private and public spheres. A cyborg resides in both simultaneously.
Our interactions with and impact on reality (our agency)	Cyborgs and bots that interact with the world in our name	The cyborg is in many ways a confidant, but it is virtually impossible for it to maintain our privacy as a confidant should

`   C) Reversal from information scarcity to information abundance

In the past, people used to believe that the reason for most human failures was a lack of knowledge or lack of information. We often hear phrases like "If only I had known X, I wouldn't have made the mistake of Y." One of the reasons for the lack of knowledge is the inability of the human brain (the natural information processor) to grasp all the knowledge and information about the world. Due to this inherent lack of knowledge, humanity has found two different solutions: one was to record its knowledge in writing, organize it, and store it in books, encyclopedias, and so on. The other was to design scientific and mathematical methods intended to generalize knowledge and define the laws of nature based on exemplars or case studies. A few prominent examples of this approach are the discovery and formulation of the laws of nature by physical and chemical experimentation, the development of statistical methods to help us make general conclusions based on samples, and the inductive approach – which can help predict the future based on past experiences, and the study of correlations between events.[428] The age of ICTs has introduced new and greatly enhanced – way beyond human – abilities for storing, analyzing and disseminating knowledge and information, causing a paradigmatic shift in the ways we obtain information and knowledge. The "encyclopedic" approach, according to which the source of knowledge is what we can learn from observing nature, is being replaced by real-time sources of knowledge gained by surfing the web and its databases.[429]

---

[428] David Hume, *A Treatise of Human Nature* (Oxford: Clarendon Press, 1896), 263-274.

[429] Levin and Tybulsky, "The Constructionist Learning Approach...", 2467.

These developments gave birth to a new mindset that Yuval Noah Harari calls "dataism" – a philosophy that perceives the universe as consisting of nothing but data flows[430] and all the biological beings, including human beings, as biological algorithms and data processing systems that thanks to ICTs can connect to a single all-encircling information network.[431] In the dataist ideology, information flow is the supreme value, and thus, the purpose of humanity is to "maximize dataflow by connecting to more and more media."[432] This supreme purpose is shared by every person connected to the network. Therefore, all things should be connected to the Internet of Things to exchange information, while the worst sin a network node could commit would be to block dataflow, to withhold information rather than spread it through the network. The freedom of information required by dataism is not the freedom of speech that stems from liberal humanism. Freedom of speech, granted to every human being, made possible the conception of privacy, which guaranteed the right to be let alone and to have a private sphere where one is autonomous to think, feel, decide and do what one wills without being punished as long as one does not violate any laws or regulations. In contrast, freedom of information is a right given to information, not to people. Information's "right" to free flow through the network is a basic need of dataism,[433] but it is a violation of people's

---

[430] See Floridi's general definition of information discussed in Chapter 8.

[431] Yuval Noah Harari, *Homo Deus: A Brief History of Tomorrow* (London: Harvill Secker, 2016).

[432] Ibid. This idea is perfectly aligned with the meme theory, which regards human beings as the vessels and transmitters of memes, whose primary purpose is to spread information and ideas.

[433] As we shall see in the discussion of privacy as order, information flow is not only a need; entropy makes it a must.

right to keep their private and personal information to themselves or to share it exclusively with confidants chosen by them. Therefore, for the purposes of this book, **freedom of information is an undermining of both general and deep personal privacy, as it undermines one's right to keep information to oneself without sharing it with others.** The "religion" of dataism is becoming the highest authority in matters considered most intimate and private. For example, because Google (or Facebook) clearly knows things about you that you may not be aware of yourself,[434] you may believe you should choose your spouse based on Tinder's recommendations rather than based on your own feelings. The Internet knows more about our health and genetic structure than any doctor – therefore, the Internet should decide which medical procedures we should undergo and when. In other words, the source of authority that tells us how we should feel and what we should do is not within us, as liberal humanism would have us believe; it is out there in the Internet network. The humanist ideal, whose biological foundation had been shaped over millions of years of evolution, is being challenged by the idea that ICTs may be able to understand us – our feelings, desires and thoughts – better than we can understand ourselves. The logical conclusion is that once the big data pools get to know people better than the people themselves, algorithms and data controllers will have the authority to make the most important decisions in our lives. According to Lévy and Pentland, **this is (or will be) the end of authentic independent individual thinking; it will be replaced by collective intelligence run by "the big brother" and voiding personal privacy.**

---

[434] See Chapter 8 for a more detailed discussion and definition of the "humps": information about you that is known to others (one or more confidants, or the public), but not known to you.

The following table summarizes the impact of the reversal from information scarcity to information abundance (or over-abundance) on the basic components of our worldview:

Main changes in the age of ICTs Basic worldview components	Reversal from information scarcity to information abundance	Resulting impact on various components of personal privacy
Our self-conception (who we are)	The network knows more about us than we know about ourselves.	The network replaces our authentic selfhood as the source of authority for our self-definition.
Our mutual interactions (how we socialize)	Other people can mine any piece of private personal information about us.	Our World 2 is exposed. Much of our general and deep privacy is breached.
Our conception of reality (our metaphysics)	Networks replace nature as the source of knowledge; however, that knowledge is fluid and has questionable validity[435].	Our private personal information is out there in the network and can be searched. The distinction between the private and public spheres is blurred.
Our interactions with and impact on reality (our agency)	The network shapes our knowledge, thoughts, and desires.	The network dictates values, ethics and norms of behavior (dos and don'ts).

---

[435] Online information constantly changes and contains a lot of fake news.

D) Blurring of the distinction between reality and virtuality

The dichotomic distinction between the physical and virtual worlds stems from the duality of body and soul – an idea that goes back to the days of ancient Greece and has been adopted by the Orthodox Christian Church.[436] In the Age of Enlightenment, it was developed further by philosophers like Descartes and Immanuel Kant in *Critique of Pure Reason*.[437] In this book, I am using Karl Popper's "Three Worlds" theory as an expansion of this dualist perception of the universe, which has become a fundamental part of Western culture but has been undermined from several directions since the second half of the twentieth century.

The blurring of the distinction between the physical and the virtual creates a new, complex reality that is neither purely physical nor purely virtual but is a little bit of both. In this new reality, the line between real and virtual is vague. The result is a growing number of virtual identities and "on-life" and "off-life" personas. One aspect of this blurring is differential ethics for online and offline behavior: for example, concealing the truth about yourself in a dating app is legitimate; committing rape or murder in a video game is not considered a crime. However, things become less clear when you consider, for example, online pedophiles. Is looking at photos of children online a crime? Seemingly, using your imagination while watching photos is part of the

---

[436] Ess, "At the Intersections Between Internet Studies...", 91.

[437] Emmanuel Kant, *Critique of Pure Reason*, ed. and trans. Paul Guyer and Allen W. Wood (Cambridge: Cambridge University Press, 1998).
In this book, Kant makes a distinction between the world of phenomena, which we know, and the "actual" world which the human mind cannot grasp. We can only perceive it through our senses, and then interpret and organize it using a priori concepts. In other words, there is a physical world that is inaccessible to the human mind, and a virtual world of human reason/cognition that can make claims about the physical world.

private sphere, which cannot and should not be supervised, lest we turn into "thought police," but norms of behavior accepted in the public sphere are penetrating into the private sphere.

The following table summarizes the impact of the blurring of distinctions between reality and virtuality on the basic components of our worldview:

Main changes in the age of ICTs	Blurring of the distinction	Resulting impact on various components of personal privacy
Basic worldview components	between reality and virtuality	
Our self-conception (who we are)	One can have multiple "selves," some of which reside solely in the virtual world.	Selfhood is multiplied and decentralized, exceeding the limits of our body and mind into multiple Internet applications.
Our mutual interactions (how we socialize)	We can interact through external virtual proxies and bots	Popper's World 2 is visible in the virtual space and thus becomes publicly exposed and ceases to be private
Our conception of reality (our metaphysics)	The world is becoming populated by immaterial, virtual entities	Some of our private personal information is placed in virtual beings we cannot control
Our interactions with and impact on reality (our agency)	Bots serve as agents and virtual proxies	Companies like Facebook and Google have become confidants in our general privacy.

	Contents	**Status**	**Examples**	**Examples related to personal privacy**
World 1	Physical objects	Objective	Data pools	Body cameras, smartwatches.
World 2	Mental experiences	Subjective	Pain, joy	Deep privacy.
World 3	Knowledge	Objective	ICTs	Deep privacy information that became exposed.

In conclusion, the age of ICTs is changing the relationship between the concepts described in Chapter 6:

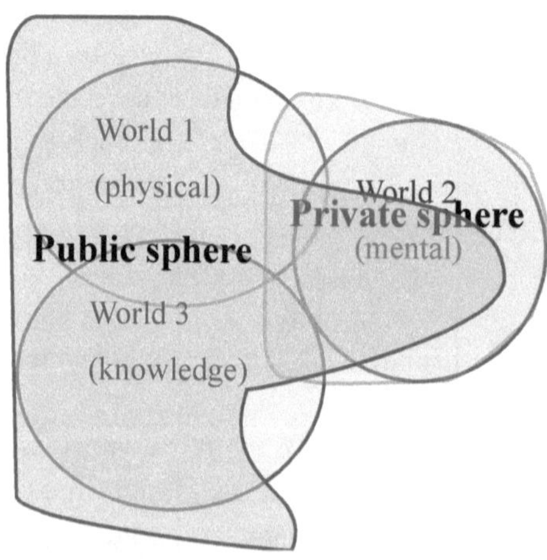

Relationship between the private and public spheres in the age of ICTs

ICTs are blurring the distinctions between physical and virtual, between human and machine, and between the private and public spheres. As a result, the public sphere is penetrating the private sphere in general, and Popper's World 2 in particular, erasing deep personal privacy that resides in World 2 and displacing personal privacy into Worlds 1 and 3.[438] I am not saying that World 2 is vanishing in the sense that we no longer think or feel, but I am saying that ICTs are exposing World 2 so that growing parts of it stop being private and intimate.

Now, let us review the state of personal privacy in the age of ICTs using the three-dimensional space model:

Values: The key values of the modern era were that every human being is an autonomous individual with specific rights and obligations that include basic human rights – the right to property, free speech, protection from violence and abuse, civil society, and the free market. The age of ICTs is challenging these values in the following ways:

The value of individualism that evolved in Western democratic culture and was founded on the unconditional natural rights to freedom, equality and justice gave birth to the Western conception of personal privacy, defining the limits of individuals' autonomy and the norms of behavior in the private and public spheres. Justice meant protecting people's natural and universal right to fulfill their individuality. In the age of ICTs, the tables have turned. Now, it is about trying to justify the individuality of people who are nodes in the network, the main argument being that when taken together, the autonomous decisions made by each of the network nodes based on partial information eventually create optimal spontaneous order. In other words, the individuality of network nodes – and, consequently, of the individuals represented by those nodes – is justified as contributing to the creation of a spontaneous order that is best for the entire network. Because "the network" in the age of ICTs

---

[438] I am not saying that world 2 is vanishing, only that it is becoming more and more exposed, and its privacy is vanishing.

demonstrates spontaneous order, which is an important natural phenomenon, "the redefinition of individuals as nodes in the network, and of those nodes as separate atoms that are flexible and adaptable to the network fluctuations, requires that they be free individuals that would allow spontaneous order to happen."[439] Thus, individuals' freedom in the free market gets three justifications: normative, scientific, and technological.

The age of ICTs introduces four major social transformations:

a. Shifting from the primacy of entities over interactions to the primacy of interactions over entities
b. Blurring of the distinctions between human and machine
c. Reversal from information scarcity to information abundance (or over-abundance)
d. Blurring of the distinction between reality and virtuality

These raise several questions that have a lot to do with personal privacy:

- Who are we, where do we end, and what are our natural rights? Does a cyborg entity that exists beyond the physical body have the same natural rights? And if so, to which parts or aspects of the cyborg do they apply?
- Does our identity include our virtual online identities, and do those have natural tights, too?[440]

---

[439] Fisher, *Capitalism be-idan ha-tikshoret ha-digitalit*, 85.
[440] What is the meaning of "natural rights" when talking about an artificial entity?

Without clear answers to these questions, there can be no clear answer regarding personal privacy – a notion that derives from a conception of the individual as having natural rights. The more specific questions concerning personal privacy are:

- Can privacy be protected if it is spread across several nodes in the network?
- Is the privacy of a virtual identity equivalent to the privacy of a human individual?
- How are a network node's connections to be treated in terms of personal privacy? Is the connection itself a part of privacy, or only the content that is exchanged?
- What is the status of data aggregated in big data pools in terms of privacy?

And above all, can personal privacy be protected at all in today's world? If it can, what would be the best ways to protect it (regulations, technologies, social norms…)?

All these issues challenge the argument that "the network" would spontaneously give people their natural rights and protect them. As we have seen, the justifications for those rights and values – and, consequently, the conception of personal privacy that is based on them – are being constantly undermined in the age of ICTs.

Norms and regulations: The legal systems and privacy protection laws and regulations have difficulty keeping up with the challenges posed by ICTs. In 1980, the OECD published eight principles that were intended to provide the normative basis (expected norms of behavior) for information privacy protection laws and regulations.[441]

---

[441] See Chapter 7, "Metaphysical Parts", for a list and discussion of the

Those are the ethical standards that require confidants to treat individuals' personal information the way confidants should. However, the large Internet companies, despite being confidants and despite their definition as service providers, do not respect the OECD principles – to the extent that Facebook founder Mark Zuckerberg openly declared in 2010 that privacy was no longer a social norm.[442] Moreover, they argue that the loss of information privacy is a price the users must pay in exchange for their "free" services. **Internet companies are playing a double game: they ask their users to provide personal information and to treat them as confidants, assuring them it is necessary in order to receive their services, yet they are financial organizations that use that private information as a resource and a commodity to trade.**

Following the publication of the OECD principles, many countries passed privacy protection laws during the 1980s,[443] but they could not predict the far-reaching consequences of the advent of ICTs, which led to a *non liquet* in this field. The EU GDPR – which was intended to provide EU citizens with a comprehensive solution for information privacy issues – has recently been challenged in court by Facebook and other Internet companies.[444]

---

eight principles.

[442] Zuboff, *The age of surveillance capitalism*, 48.
Other executives of large Internet companies have made similar claims. See, for example:
Gottlieb, "Privacy: A Concept Whose Time Has Come and Gone", 156.

[443] E.g., the Israeli 1981 Privacy Protection Law.

[444] Chris Fox, "Google and Facebook accused of breaking GDPR laws", *BBC Online*, 25 May 2018 https://www.bbc.com/news/technology-44252327 (Accessed 12/5/2024)
Neil Hodge, "Ireland GDPR caseload nearly doubled in 2019",

Knowledge: Scientific, mathematical and engineering knowledge is advancing rapidly. The field of knowledge that is relevant to this book is the information and communication technologies described in Chapter 1, which have an unprecedented ability to peep into Popper's World 2. For example, Facebook has the ability to detect your sexual orientation before even you are aware of it.[445]

Spiritual culture: The philosophy of the network society defines individuals as nodes in the network.[446] Information is regarded as the primary economic and cultural resource. Neo-liberalism – as opposed to liberalism – sees the individual as a tool to serve economic and social ends rather than a value in itself.[447]

In relation to privacy: The distinction between the private and public spheres is increasingly blurring. Individuals are nodes in the network, and as such, their interactions are becoming more important than their individuality. Moreover, as mentioned in the section on self-conception and the "death of individuality," one's personality can be reshaped by network pressure. On the other hand, ICTs are opening up

---

Compliance Week, 20 February 2020 https://www.complianceweek.com/data-privacy/ireland-gdpr-caseload-nearly-doubled-in-2019/28475.article (Accessed 12/5/2024)

[445] Nikhil Bhattasali, Esha Mati, "Machine 'Gaydar': Using Facebook Profiles to Predict Sexual Orientation", Course paper, CS 229 – "Machine Learning", Department of Computer Science, Stanford University, Fall 2015. https://cs229.stanford.edu/proj2015/019_report.pdf (Accessed May 2024)

[446] See Chapter 3.

[447] Fisher, *Capitalism be-idan ha-tikshoret ha-digitalit,* 192: "Human beings have become means – a sharp turn from the humanism of the enlightenment, which saw human beings as having intrinsic value which cannot be quantified or replaced, subjects, and their own ends."

new possibilities for constructing different identities, including gender identities. As a result, all the components of personal privacy are being challenged.

Social culture: The network culture is characterized by **network production** – a non-hierarchical, decentralized production system where one is a prosumer (producer and consumer at the same time), and one's experience and personal privacy are commodities traded on the knowledge market; **network work** – with no clear distinction between work time and leisure time, between home space and workplace space; **the network market** – a market whose targets are not anonymous mass consumers, but niches and even specific individuals ("long tail" strategy); and **the network human** – each person is an atomic node in the network, flexible and adaptable to the network's fluctuations. This social culture undermines conventional **personal privacy**. The laws and regulations designed to protect our privacy fail to cope with the challenges posed by ICTs. Social norms aimed at protecting personal privacy, such as the confidentiality of one's private writings, are no longer binding. This is related to the four key transformations described in this chapter.

Technological culture: The technological culture of our time is defined by information and communication technologies (ICTs) and their impact on personal privacy. The ICTs' operation mode is using information as a resource. In Chapter 8 – Personal Privacy Information – I showed how the discussion of personal privacy may be replaced with a discussion of personal privacy information.

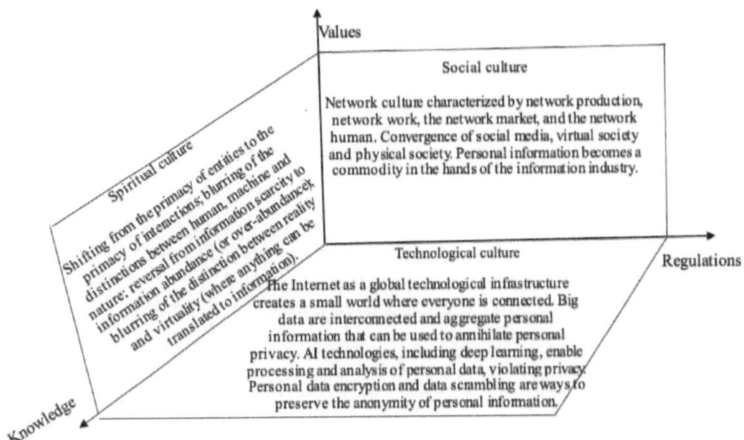

Personal privacy in the age of ICTs, according to the three-dimensional space model[448]

---

[448] Levin, "Cultural trends in a digital society", 13-21.

A Metric for Intercultural Comparison of Personal Privacy

As described in Chapter 3, any given culture (society) can be represented on a three-dimensional chart[449] whose axes are values, regulations and knowledge:

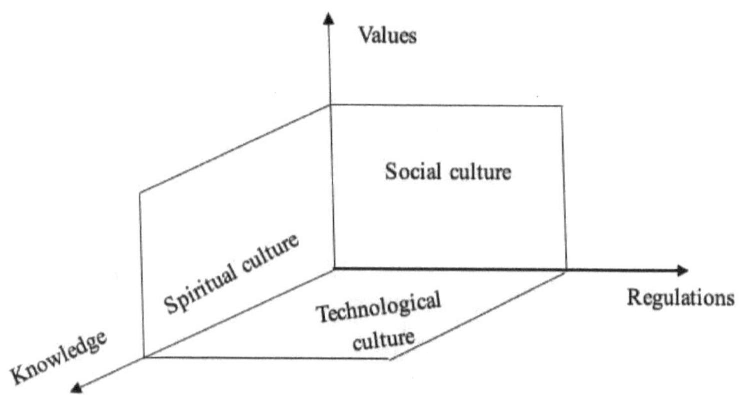

Throughout this book, I have been using this model to describe the state of personal privacy in the various cultures I reviewed. Now, I would like to add a metric for calculating and grading the extent of privacy allowed by each particular culture in order to make intercultural comparisons.

My metric uses the three-dimensional space model of culture and represents personal privacy (PP) **as a cube whose volume is PP = V · R · K:**

- V – values;
- R – norms and regulations;

---

449 Ibid.

- K – knowledge;
- Value range {0…10}

Values and regulations have linear influence on personal privacy: the greater they are in amount and power, the better personal privacy is protected. This can be seen from the section on individualism and the enlightenment values of selfhood.

In contrast, there is no linear association between personal privacy and knowledge. While in the past, increased knowledge was associated with a greater extent of personal privacy, in the age of ICTs, personal privacy is reduced by publicly exposing personal private information in a variety of ways.[450] This non-linear association can be represented by the following **function**: $K' = -0.5K^2 + 5K$ (K standing for particular knowledge by a given culture).[451]

This function monotonically increases for $K \leq 5$, monotonically decreases for $K > 5$, and is zero for $K = 0$ and $K = 10$. A state of $K = 0$ means the culture has no knowledge at all, which is impossible because at least some knowledge is a necessary a priori condition for the existence of any culture whatsoever. A state of $K = 10$ means the culture has all the knowledge that can effectively void personal privacy; that is also impossible, at least regarding deep personal privacy, which resides in Popper world 2, because our inner world is continuously and rapidly repopulated with thoughts, feelings etc. which cannot all be exposed at

---

[450] See a detailed discussion in the earlier sections of this chapter.

[451] Other functions may also be used to represent the influence of knowledge on privacy, as long as they preserve the non-linear association between the two.

any given moment. Therefore, the theoretical limits of our information function cannot exist in the real world.

Let us now use this metric to compare the extent of personal privacy in all the previously reviewed cultures:

- Personal privacy in primitive cultures
- Personal privacy in antiquity
- Personal privacy in the modern era
- Personal privacy in the age of ICTs 1.0
- Personal privacy in the age of ICTs 2.0

Each culture will receive a score based on the extent of personal privacy that it allows:[452]

1. Personal privacy in primitive cultures was given the following values: $V = 2$, $R = 2$, $K = 1$, making $K' = 4.5$; therefore, $PP = 2 \cdot 2 \cdot 4.5 = 18$.
2. Personal privacy in antiquity was given the following values: $V = 6$, $R = 6$, $K = 3$, making $K' = 10.5$; therefore, $PP = 6 \cdot 6 \cdot 10.5 = 378$.
3. Personal privacy in the modern era was given the following values: $V = 8$, $R = 7$, $K = 6$, making $K' = 12$; therefore, $PP = 8 \cdot 7 \cdot 12 = 672$.
4. Personal privacy in the age of ICTs 1.0 was given the following values: $V = 8$, $R = 8$, $K = 8$, making $K' = 8$; therefore, $PP = 8 \cdot 8 \cdot 8 = 512$.

---

[452] The values given to each parameter faithfully represent its relative strength in each particular culture, according to the descriptions in the previous chapters of this book.

5. Personal privacy in the age of ICTs 2.0 was given the following values: $V = 8$, $R = 8$, $K = 9$, making $K' = 4.5$; therefore, $PP = 8 \cdot 8 \cdot 4.5 = 324$.

Let us display the obtained scores on a chart:

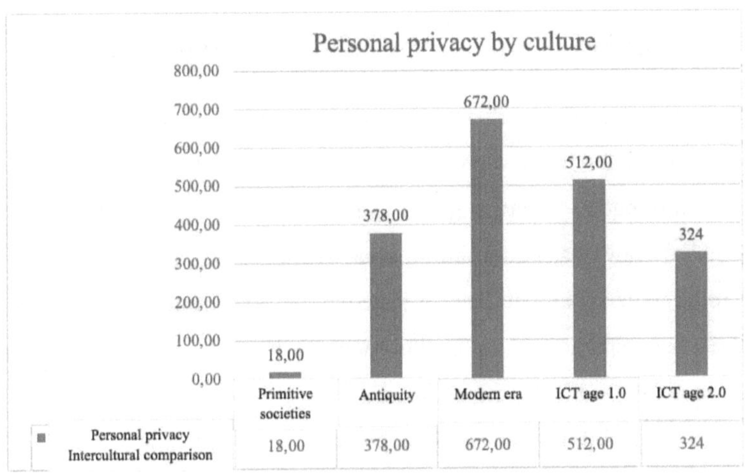

Chapter 10

Privacy Information is Bound to Be Dispersed

What Makes Personal Privacy Information Dispersible

Protecting personal privacy information against public exposure is challenging for two reasons: one, formulated by John Perry Barlow, is that information behaves like a life form;[453] the other, explained by Richard Dawkins, is that information elements are memes, which are similar to genes in that their ultimate goal is to replicate.[454] Barlow lists three key characteristics of information that explain why personal privacy information is dispersible by nature:[455]

(1) Information is an activity: it is a verb, not a noun; it is experienced rather than possessed, and it has to move (flow). Information that isn't moving ceases to be information or becomes frozen until it can move again. In regard to personal privacy information, information that an individual keeps to themself and that cannot be passed on to other entities – even potentially – becomes irrelevant and

---

[453] John Perry Barlow, "The Economy of Ideas", *Wired Digital, Inc.*, 01/1994, 1-15.
https://www.wired.com/1994/03/economy-ideas/ (Accessed 22/8/2020)
[454] Richard Dawkins, *The Selfish Gene* (Oxford: Oxford University Press, 1976).
[455] Barlow, "The Economy of Ideas".

meaningless. For example, if you know you are sensitive to a certain smell, but nobody else knows about it, and your behavior does not give it away, this information is irrelevant to your interaction with the world; it is meaningless, and therefore, it does not meet the third criterion of the GDI.[456] The way in which information spreads is very different from the distribution of physical goods: it can be transferred without leaving the possession of the original owner. It can replicate itself endlessly and spread without falling apart into little pieces. **However, even though the sender of the information does not lose it, they do lose some power (knowledge is power), while the power of the information receiver increases respectively.**

In the age of ICTs, information in general and personal privacy information in particular is dispersed via physical and virtual platforms (the Internet and social media). Personal privacy information is being aggregated continuously, with virtually no filtering, and often "sits" until it is needed. I will again mention Helen Nissenbaum and the connection she makes between information and personal privacy by defining personal privacy as a flow of information.[457] Personal privacy information flows from the sphere of deep personal privacy into the sphere of general personal privacy and onto the public sphere.

(2) Information is a life form: This argument is based on meme theory, which was introduced by Dawkins in his book *The Selfish Gene*.[458] The Oxford English Dictionary defines a meme as "a cultural element whose transmission can occur by non-genetic means, especially

---

[456] This is logically similar to the philosophical argument that a falling tree does not make a sound if there is nobody there to hear it.

[457] Nissenbaum, *Privacy in Context...*, 127.

[458] Dawkins, *The Selfish Gene*, 245-260.

by imitation."[459] According to Karl Popper's theory, World 3 is populated with knowledge and ideas; therefore, I will also place the majority of information in World 3: any information that is shared by at least two people resides in World 3, while deep personal privacy information – as explained in previous chapters – resides in world 2 Because information "wants" to flow, as argued by Barlow, every piece of information is also a meme which wants to replicate and spread. This is true of any information, including personal privacy information. For example, when our body "wants" to spread information about itself, it sends out messages that can be perceived by the senses (e.g., sight, smell, touch) or even messages from which our mental state or traits can be inferred.[460] Dissemination of personal privacy information has increased greatly in the age of ICTs. Governments and commercial organizations aggregate pieces of personal privacy information as described above, analyze them and share them for commercial and other use. The information is usually replicated many times and stored simultaneously in multiple databases (e.g., every Internet app stores personal information for its own and others' needs), and so the memes replicate and spread, violating both general and deep personal privacy. Other mechanisms, such as entropy and human curiosity, further increase the dispersion of personal privacy information.

(3) Information is a relationship: Information acquires meaning only when it flows from its source to its recipient. This is the process through which data turns into meaningful information. Further meaning is added by the way the information is stored and presented. For example,

---

[459] https://www.oed.com/search/dictionary/?scope=Entries&q=meme (Accessed May 22, 2024)

[460] Some claim, for example, that fear has a smell. The body communicates information using a World 1 indicator (scent).

a piece of information like your bank account balance has meaning in itself at a given time point, but when presented in a series of your account balances over a period of time, it can give a much better picture of the state of your financials, whether it has improved or worsened, and where you stand financially compared to other people. Data also acquire meaning through our interactions with ourselves, particularly if we have multiple online and offline identities (whether they exist alongside each other or replace one another). Obviously, when private personal information flows (or has the potential to flow) from the private sphere to the public sphere, it affects the privacy of the information subject.

Personal Privacy Information as "Order"

First of all, let us clarify: What **is order?** In the most general sense, the order in any system (whether physical, social or biological) is when its various parts are arranged according to a particular sequence, pattern, or method in relation to each other – "a law, or a principle, that governs the relations among the parts."[461] Order can be artificial, i.e., human-made, such as the lexicographical order of books in a library, or it can be natural – physical, biological etc. – in which case it is referred to as "a law of nature". Either way, the order has three main characteristics: complexity, lawfulness, and quantitative.[462]

The term "order" applies to the relationships between the various parts of the system and to the ordering principle that governs them. It can be both internal – defining the arrangement of the parts in relation to each other, and external – defining the relationships between

---

[461] Ruth Lorand, *Aesthetic Order: A Philosophy of Order, Beauty and Art* (London and New York: Routledge, 2000), 9.
[462] Ibid., 9-19.

the system parts and entities that are not part of the system. The order may change as parts are added or subtracted, but that does not change the ordering principles (laws) of the system. Such laws can be absolute, like the laws of physics above atomic level, or statistical, like in quantum physics. Without ordering principles, there will be chaos (disorder). The degree of order means the degree of a system's conformity to its ordering principles, and so we can define a system as "more ordered" or "less ordered." In other words, the degree of order is the gap between the ideal picture of the world as it should be and the actual state of events. Order and disorder are not a dichotomy;[463] most systems are somewhere on the spectrum of conformity between complete order (in physics – zero entropy at absolute zero; in economic and social systems – complete equilibrium) and complete chaos (when the system's ordering principles no longer apply and it bounces from state to state without any ability to reach equilibrium). An information system is in total order (zero entropy) if zero information flows in, out, or through it.

Is personal privacy "order"? To understand whether personal privacy can be defined as "order," let us examine it through the three characteristics of order: complexity, lawfulness, and quantitative. Two aspects of the **complexity of personal privacy** have been discussed throughout this book: one aspect is the relationship between every individual's agency and society, which is comprised of different reference groups and other individuals; the other aspect is the multiple basic components extracted from the various definitions of privacy, which carry information and together make up all our personal (and

---

[463] Efron Razi and Pinhas Yehezkeally, *Ha-olam eino liniari: Torat ha-maarakhot ha-merukavot – gorem khadash be-nihul* [Real Life Is Not Linear: Introduction to Complex System Theory] (in Hebrew) (Tel Aviv: Ministry of Defense Publications, 2007), 68-71.

private personal) information,[464] and the interactions between them. The **lawfulness of personal privacy** is the system of laws, regulations and social norms designed to protect personal privacy and personal information, which is an integral part of personal privacy. The **quantitativity of personal privacy** has been discussed in Chapter 2. As we can see, personal privacy meets the criteria of order, and therefore, it is possible to discuss the existence of personal privacy as order and the degree of order (or disorder) in personal privacy in the age of ICTs.

The Order in Personal Privacy in the Age of ICTs

The age of ICTs has personal privacy in a state of turbulence.[465] Though it is a state between complete order and complete chaos, it is not a point on the spectrum of conformity but a whole different range of states characterized by "whirlpools" that disrupt the existing order of things (like chaotic currents that disrupt the water flow). This is a state typical of social and political systems on the verge of a revolution when the old order and its agents fail to efficiently manage the system and the events, but a new order has not yet formed. The age of ICTs is a revolution that will result in the rise of a new civilization – the network society, whose primary resource is information. As the new network society is taking shape, there are several whirlpools, all of which involve the use of ICTs and information as their primary resource.

---

[464] Personal information that isn't private means information about us that is also known to other people, while private personal information is information about us that is possessed or controlled by us.

[465] Razi and Yehezkeally, *Ha-olam eino liniari…*, 71-72.

A physical state in which a system's behavior is unpredictable even though all its ordering principles are known.

The political whirlpool: Shaking of representative democracy as the dominant political regime in liberal Western societies, enhanced by such uses (or abuses) or ICTs as Donald Trump's practice of political management by Twitter above the heads of the traditional branches of the government; the Facebook-Cambridge Analytica data scandal, which is an example of mind engineering by means of AI tools and social media; cyber attempts (probably Russian) to interfere in US and French elections; the NSA's improper use of ICTs to violate human liberties, as appears from the documents leaked out by Edward Snowden; etc.

The economic whirlpool: The emergence of surveillance capitalism, targeted advertising, blurring of the distinctions between producers and consumers, niche production, etc.

The social whirlpool: Widespread use of virtual identities, loss of autonomy and anonymity, blurring of the distinctions between the private and public spheres, the primacy of interactions over entities, and other phenomena described in the previous chapters.

All these whirlpools are challenging the privacy order that existed in pre-network society. The shaking of the order can be felt in every aspect of our private lives as well as in the legislation, norms and ethics regarding personal privacy. The laws of the economy seem to be changing, and it is not yet clear whether the principles and rules of capitalism still stand or whether we are seeing the rise of a new economic system. The Fordist social order, which guaranteed social welfare and economic growth through a system of balances between the government, the producers and the workers, is being challenged as well. **Each of the whirlpools has an enormous impact on personal privacy.**

Here are some examples of the turbulence in the field of personal privacy:

- "Victims of cybercrime suffer great emotional, physical and financial damage. Studies have shown that the effects of cybercrime may be similar to those of burglary."[466] Cybercrime constitutes unauthorized penetration of our private space and causes great emotional distress, shaking the foundations of privacy, particularly the two basic components of mind and private space.

- In Israel, the time of the COVID-19 pandemic was marked by numerous invasions of personal privacy.[467] The Shin Bet used its location app to monitor citizens, and "outrageous ideas such as giving every citizen a 'contagiosity score' to reflect his or her level of risk were proposed. It was the year when the Ministry of Communications requested that cellular companies provide identified data about their subscribers, including intimate data. The right to privacy is protected by the Basic Law of Human Dignity and Liberty, but the gap between the constitutional status of this right and its actual abuse by the government is

---

[466] Asaf Amir, "Kshe ha-pratiyut shelanu mekhulelet ba-reshet: al ha-hashpaot ha-psychologiyot shel pish'ey ha-cyber" [When our privacy is violated online: the psychological effects of cybercrime] (in Hebrew), *Anashim u Makhshevim*, 10 February, 2021 https://www.pc.co.il/featured/331923/ (Accessed 11/5/2021)

[467] Tal Shakhaf, "2020 hayta shana shkhora la-pratiyut be-Israel" [2020 was a dark year for privacy in Israel] (in Hebrew), *Ynet*, 29 January, 2021 https://www.ynet.co.il/digital/technology/article/ryAzJe11xu (Accessed 11/5/2021)

jaw-dropping."[468] This is a great example of both the political and social whirlpools, with direct violation of personal privacy and privacy of personal information – particularly the anonymity component – and disclosure of private information.

- A 2020 leak from the Elector app, reported by the Calcalist newspaper, led to the exposure of the personal details of 6.5 million Israeli voters, including names, ID numbers, addresses, phone numbers, and information collected by the Likud party activists such as age, communication language, health issues and political affiliation.[469] This is another example of turbulence in the political system, which feels free to use people's private personal information. Israel's Privacy Protection Authority concluded the investigation by ruling that "Elector Software and the political parties Likud and Israel Beiteinu violated the Privacy Protection Law and the regulations that stem from it."[470]

- Legal systems are in turbulence when it comes to personal privacy, unable to adequately address the challenges posed by ICTs. The existing legal attempts to address those challenges are either too vague or unenforceable. For example, to protect yourself from identity theft, all you can do is be cautious and closely monitor the things you care about.

---

[468] Ibid.

[469] Ibid.

[470] Ibid.

- The Israeli Ministry of Health disclosed residents' personal information to Pfizer in return for COVID-19 vaccines: "The 2020 contract between the Israeli government and Pfizer, which supplied Israel with millions of COVID-19 vaccines in exchange for complete information about the vaccine receivers, drew the public's attention to the vast amounts of information the Ministry of Health has about us – sensitive information that includes every health condition and every treatment we've ever received. The Ministry of Health created a database of COVID-19 patients, which includes their complete medical history. A few disturbing questions arose along the way: Was the extensive vaccination campaign actually a large-scale clinical trial in which we all participated without our consent? And how can we know how our medical information will be used, who owns it, and who owns the profits it can yield?"[471] This is an example of how big data technology contributes to the turbulence in personal privacy information.

All these case studies reflect the turbulence of the system of personal privacy. The debate concerning it revolves around the questions of consent to lose control over our personal privacy information and control over the flow of personal privacy information: the direction of the flow – from the private sphere to the public sphere; the intensity of the flow – how much of our personal privacy information is disclosed to

---

[471] Ibid.

other entities; and the ways personal privacy information is spread through the Internet by means of ICTs.

The Degree of Disorder in Personal Privacy

This section will introduce a quantitative metric of the degree of disorder in personal privacy. Ralph Stacey defines five parameters for placing a complex system on the spectrum between order and chaos:[472]

- The intensity of information flow between the nodes of the system: Tremendous amounts of personal information flow through the Internet every day, which puts the system of personal privacy and personal privacy information high on the spectrum of disorder.[473]

- Personal information as the primary resource of the ICT culture is being collected, aggregated, analyzed and disseminated using a wide range of means, including the IoT, AI systems, smartphones, communication networks, and above all – sophisticated apps they employ all of those. Therefore,

---

[472] Razi and Yehezkeally, *Ha-olam eino liniari…*, 174.

[473] From the 2017 Database Registrar Report by then Commissioner of the Privacy Protection Authority Adv. Alon Bachar, November 2018, Ministry of Justice – Privacy Protection Authority, pp. 5-6: "The extensive amount of personal information that is being collected includes our interests, daily activities, desires, ambitions, preferences, dreams, and the people we are in contact with. All this information is being systematically and efficiently processed and aggregated to create detailed, segmented personalized profiles".

for this parameter, the system of personal privacy is also placed high on the spectrum of disorder.[474]

- The level of interaction between the nodes of the system and between the system and the world around it: One of the primary components of the network society is the network human, who is simultaneously a node in the social network and an agent who interacts with the network and affects it. Every individual who is a node in the network extensively interacts with other nodes, exchanging personal and private personal information. This activity shapes the state of personal privacy. Furthermore, not only are individuals themselves nodes in the network, but their personal information is contained in multiple devices (such as smartphones, databases, PCs and IoT devices), which are nodes in the network, too, and they are all connected to each other through the Internet, constantly interacting, exchanging and updating

---

[474] Ibid.: "Every area of our daily life is associated with sophisticated technologies and increased exposure of our information online. Living in the digital age means our every action, or inaction, leaves a unique digital footprint which includes information about every place – private or public – we have visited, and every website or app we have used – at home, at work, on the street, from our mobile phone, smart watch, TV set, car, smart toy, etc. [...] All this information is being systematically and efficiently processed and aggregated to create detailed, segmented personalized profiles which are used, and even sold to third parties, for a variety of commercial, political and other purposes. The information is often collected without our full knowledge and without our free consent."

personal information. Thus, this parameter shows a high degree of disorder, too.

- The system nodes' motivation and involvement level: Because personal information is the primary source of income for large Internet companies, they have a very strong motivation to keep using it and are doing everything they can to remove any legal or normative obstacle that might prevent them from doing so. Governments and other non-commercial organizations also have a motivation to control personal information, as it increases their control and management capabilities. Thus, the fourth parameter also shows a high degree of disorder.

- The system's hierarchical structure: In contrast to the Fordist era, organizations (whether commercial, political or social) in the network society are typically flat because the social and technological network allows direct contact between nodes with little need for middlepersons. This leads to the "small world problem" with a maximum of six degrees of separation. This last parameter also puts the system of personal privacy high on the spectrum of disorder.

As we can see, in the age of ICTs, personal privacy as a complex system has a very high degree of disorder. This creates a feeling of chaos, which is intensified by declarations of the death of privacy or of its being unneeded and unjustified, and makes government (and non-government) regulation in this field a matter of urgent priority.

Entropy as the Physical Basis of the Turbulence in Personal Privacy

In physics and biology, **entropy** is defined as the **degree of disorder (chaos) in a complex system**. American mathematician Norbert Wiener, who is considered the founder of cybernetics, called it "nature's tendency to degrade the organized and destroy the meaningful,"[475] with a characteristic tendency to increase. In the specific context of information, he defined the message as "a sequence of events in time which, though in itself has a certain contingency, strives to hold back nature's tendency toward disorder."[476]

Order in personal privacy information means a clear definition of the legal regulations and social norms that govern it and of the entities who should own and control it. These determine the degree of order in personal privacy. An increase in entropy means a decrease in the degree of that order. The process of personal information leaking out of the private sphere into the public sphere may be compared to molecules of gas flowing from one container to another until they reach equal pressure, in accordance with the second law of thermodynamics.[477] You

---

[475] Norbert Wiener, *The Human Use of Human Beings: Cybernetics and Society* (New York: Da Capo Press, 1954), 158.

[476] Ibid., 27.

[477] Ran Tivoni, "Entropiya neged evolutzia: Seder ba i-seder" [Entropy against evolution: Putting disorder in order] (in Hebrew), Davidson Institute of Science Education – The Educational Arm of The Weizmann Institute of Science, July 5, 2014
https://davidson.weizmann.ac.il/online/maagarmada/doubt/%D7%90%D7%A0%D7%98%D7%A8%D7%95%D7%A4%D7%99%D7%94-%D7%A0%D7%92%D7%93-%D7%90%D7%91%D7%95%D7%9C%D7%95%D7%A6%D7%99%D7%94-%D7%A2%D7%95%D7%A9%D7%99%D7%9D-

can imagine the private and public spheres as communicating vessels connected through multiple Internet channels, each containing pieces of information. Due to the properties of information, when it flows from one sphere (vessel) to the other, it replicates itself without actually leaving the original vessel. As entropy grows, the system will strive toward equilibrium between the vessels and the categories of information (memes) each contains.

The process can be illustrated as follows:

%D7%A1%D7%93%D7%A8-%D7%91%D7%90%D7%99-%D7%A1%D7%93%D7%A8 (Accessed 18/5/2021):

"The first person to describe entropy as disorder was the 19th century Austrian scientist Ludwig Boltzmann, who looked at it from a statistical perspective. According to Boltzmann, the entropy of a particular state (e.g., organized or messy room) is determined by its probability: the greater the probability (chance) of it happening, the greater its entropy. Boltzmann claimed that the probability of something happening increases in direct proportion to its degree of disorder; in other words, systems left to spontaneous evolution will eventually reach maximum disorder. Because the connection between molecules and energy was not understood as well in Boltzmann's time as it is now, his description of entropy is based on what is visible to the eye. For example, experiments with various gases show how the gas molecules spread when a piston is pushed or a balloon is inflated. These and other experiments inspired Boltzmann and many others, up to our time, to describe entropy as "disorder"."

Stage 1:

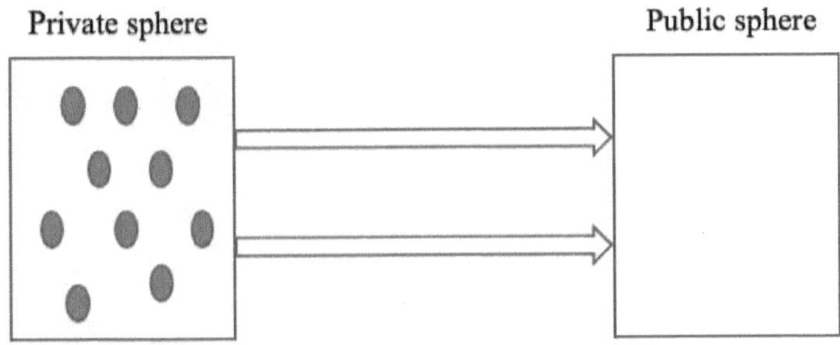

Stage 2:

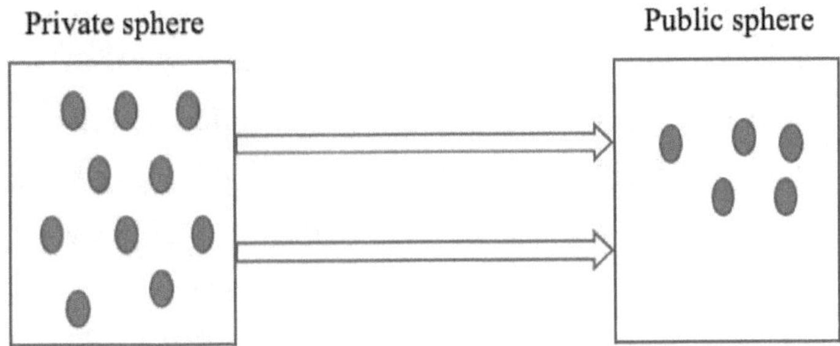

Stage 3: Equilibrium (a theoretical state of complete absence of priv

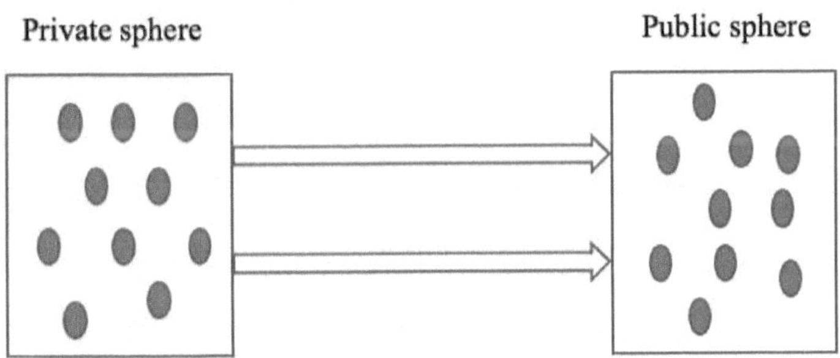

For the sake of this discussion, I will use Claude Shannon's definition of information[478] as a collection of facts about a given situation that "can lead to an increase in understanding and decrease in uncertainty."[479] Since personal privacy can be translated to information, the more information others have about a specific aspect of your personal privacy, the less uncertainty regarding your personal privacy they have. When personal privacy information leaks out of the private sphere into the public sphere, your level of privacy is decreased. Knowledge about deep personal privacy is always indirect; only those elements of privacy that are manifested in Popper's Worlds 1 and 3 can be tested and measured, so there is always some uncertainty regarding the interpretation and association between the manifestations of personal privacy that are observable in the visible world, and personal privacy itself. To measure that uncertainty, there must be a tool to assess how close one is to the actual personal privacy.[480] The uncertainty is personal privacy because it is precisely that gap between what we know about ourselves and what the world knows about us.[481] Any measurement of uncertainty must be based on the following principles:[482]

- If the outcome of an event is 100% predictable, the uncertainty regarding that event is zero. In the context of privacy, if, for example, blood pressure measurement could predict with 100% certainty

---

[478] Shannon, "A Mathematical Theory of Communication".

[479] www.businessdictionary.com/definition/information.html (Accessed 16/06/2021)

[480] See the metric proposed in Chapter 2 for calculating the extent of non-knowledge of personal privacy information.

[481] See Chapter 8, definitions of privacy and deep privacy.

[482] Rokach and Maimon, *Data Mining with Decision Trees*, 1-52.

whether the subject is feeling fear, the level of privacy regarding the feeling of fear would be zero – neither general nor deep privacy.

- The more potential outcomes an event has, the greater the uncertainty regarding that event. In the context of privacy, if those manifestations of privacy that are observable in Worlds 1 and 3 may be interpreted in more than one way (e.g. if the said blood pressure measurement might indicate one of several emotions), the level of uncertainty grows.

- If each of the potential outcomes of an event has equal probability, the uncertainty is at maximum. In our example regarding privacy, if the result of the blood pressure measurement indicates anger, fear, or embarrassment with equal probability, the uncertainty of the measurement will be high, and the privacy of the subject's emotions will be maintained.

- If there is a high probability of tracing back an element of private personal information that had flowed into the public sphere, there is a high level of certainty regarding the violation of personal privacy.

Based on these principles of uncertainty measurement, I will now show how entropy can be a tool for measuring uncertainty, i.e., privacy (as a high level of uncertainty means a high level of privacy). To do that, I have to briefly explain the mathematics of entropy and its use in the specific context of personal privacy.[483]

---

[483] For a more complete explanation of the mathematics of entropy, see

A Mathematical Formulation of the Association between Personal Privacy and the Observable Worlds 1 and 3

The definition of entropy is as follows:

$$H(X) = -\sum p(x) log\ (p(x))\ ^{484}$$

With X being a discrete random variable, x being a value of X, and $p(x)$ being the probability of getting the value x for X.

General interpretation: The sum of discrete random variables X (distributed according to p) is calculated using the entropy function H.

Interpretation in this book: X represents discrete random variables present in Worlds 1 and/or 3 whose uncertainty regarding the private sphere is calculated using the entropy function H.

Let us review the properties of the entropy function to make sure it meets the requirements for measuring uncertainty, as described above:[485]

- $H(X) = 0$ if, and only if, $p(x) = 0$ for every x except one in the relevant range. In other words, the outcomes are deterministic.
- $H(X) \leq log\ (number\ of\ possible\ outcomes)$
- $H(X) \geq 0$ because $log(p(x)) \leq 0$ for $p(x) \leq 1$
- $H(X)$ takes a maximum value if $p(x) = p(y)$ for every x, y

---

Appendix C.

[484] The base for log is not very important here.

[485] The full mathematical development can be found in Shannon's article. Here, I am quoting the formulation used by Prof. Oded Maimon in his course "Knowledge Discovery and Data Mining", given in Tel Aviv University, Spring 2018.

- If all the possible outcomes have equal probability, $H(X)$ is monotonically increasing as a function of the number of possible outcomes.

From these stem the following formulae:

For every two variables, X, Y, the cumulative information $I(X,Y)$ is the level of certainty regarding X based on the observation of Y. Formally: $I(X,Y) = H(X) - H(X \mid Y)$. The cumulative information has the following properties:

1. Symmetry: $I(X,Y) = I(Y,X) = H(X \mid Y)$
2. Non-negativity: $I(X,Y) \geq 0$
3. Maximum value: if $Y = f(x)$, then $I(X,Y) = H(Y)$
4. Minimum value: If X and Y are independent of each other, then $I(X,Y) = 0$

As we can see, the entropy function qualifies as a mathematical tool for the measurement of uncertainty, and for the purposes of this book, it can be the basis of a metric for measuring the dispersal of personal privacy information.

The following is a numerical example of personal privacy information dispersal:

Let us describe the certainty of personal information dispersal as follows: X is a collection of elements of personal information, such as marital status, phone number, age, height, weight, and other elements of information included in the basic components of privacy.[486] $p(x)$ is the probability that information element x resides entirely in the private sphere. Let us assume that $p(x) = 0.67$; therefore, $1 - p(x) = 0.33$.

---

[486] As defined in Chapters 6 and 8.

This makes the entropy of x's dispersal between the two "vessels" (i.e., the private and public spheres):

$$H(X) = - [0.67log_2(0.67) + 0.33log_2(0.33)] = 0.9149$$

Thus, the certainty $I(X)$ of the dispersal of x in the public sphere is:

$$I(X) = 1 - H(X) = 1 - 0.9149 = 0.0851$$

If ICTs are increasing the probability of x leaking out into the public sphere so that $p(x) \geq 0.85$, the entropy will be:

$$H(X) = - [0.85log_2(0.85) + 0.15log_2(0.15)] = 0.6098$$

Which means the certainty $I(X)$ of the dispersal of x in the public sphere will be:

$$I(X) = 1 - H(X) = 1 - 0.6098 = 0.3902$$

Thus, the increasing entropy increases the certainty of personal information dispersal into the public sphere, reducing the level of personal privacy. ICTs are enhancing the process of personal information dispersal from the private sphere into the public sphere.

Formally, the entropy function $H(X)$ is monotonically increasing, which means personal privacy is decreasing. The theoretical lower limit means all the private information is out in the public sphere, i.e., when $H(X) = 1$, then $I(X) = 1 - H(X) = 1 - 1 = 0$. This limit is unreachable because, at the very least, deep personal privacy can never disappear.

The entropy function as a measurement of deep privacy: Let Y be an element of deep privacy (e.g., an emotion, thought or belief). Y obviously resides in World 2, and therefore, an outside observer can seemingly know nothing about it. Therefore, let $G(Y)$ be the privacy function. However, if the outside observer has some indicators (X) observable in Worlds 1 and 3 that may, with some probability, point to Y, we can define the entropy of Y given X as follows:

$$G(Y) = H(Y \mid X)$$

with x being an indicator observable in Worlds 1 and 3 and $p(x)$ being the probability of x pointing to the presence of Y. Thus, we can use the entropy function to assess the certainty of Y or, more precisely, the certainty of knowing that Y is part of a person's World 2, in violation of that person's privacy.

To base this argument, I shall first demonstrate the association between the entropy function and the information addition gained by means of indicators or the positive change in the probability of association between the observable indicators and the privacy element we seek to uncover (a feeling, let us say, embarrassment). To show the mathematical association between the addition of indicators and the decrease in the uncertainty of a privacy element, let us define the properties of conditional entropy:[487]

$$H(Y \mid X) = \sum_x p(x) \cdot H(Y \mid X = x) = -\sum_x p(x) \cdot \sum_y p(y|x) \cdot log(p(y|x))$$
$$= -\sum_x \sum_y p(x,y) \cdot log(p(y|x))$$

- X and Y are random discrete variables
- $p(x,y)$ is the joint probability of x and y
- $p(y \mid x)$ is the conditional probability of y given x

The uncertainty regarding Y will always be equal to or greater than the uncertainty regarding Y given X: $H(Y \mid X) \leq H(Y)$. Therefore, given information addition (indicator) X, the uncertainty regarding Y is reduced to:

$$I(X;Y) = H(Y) - H(Y/X) = H(Y) - \sum_{x,y} p(x,y) \cdot log\left(\frac{p(y|x)}{p(y)}\right)$$

For the purposes of this book, The privacy of Y is reduced given indicator X.

---

[487] The mathematics is based on: Rokach and Maimon, *Data Mining with Decision Trees*, 1-52.

Now, we will see how this math is used by data mining systems analyzing big data pools to uncover deep privacy information. This is a real-life illustration of the way ICTs – which are part of Popper's World 3 – interacting with data from World 1 (the physical world) can contribute to the exposure of World 2 (the mental world, which includes people's inner thoughts, feelings and other mental and emotional elements), which is the sphere of deep privacy that should be inaccessible to others. Here is a numerical example:

Let us assume we wish to know if a person is feeling embarrassed. Embarrassment is clearly an emotion that resides in World 2 (one's mental world), and as such, falls under the definition of deep privacy, so long as the embarrassed person has not shared their feeling with others. However, we, as observers, are able to notice some symptoms belonging to Worlds 1 and 3, which may suggest embarrassment (or lack thereof) in a person under certain circumstances. Let us assume, then (without loss of generality), that blushing suggests embarrassment with some probability.

Now, let there be a group of 10 people about whom we have the following information (as per the table below):

Embarrassment \ Blushing			Blushed		Not blushed	
		Blushing	Amount	Probability	Amount	Probability
			6	0.6	4	0.4
			Conditional probabilities			
Embarrassed	Amount	Probability	4	0.8	1	0.2
	5	0.5				
Not embarrassed	5	0.5	2	0.4	3	0.6

If we pick one person from this group at random, there will be a 50% chance that they are embarrassed and, independently, a 60% chance that they are blushing. If we pick a blushed person, the chance of them being embarrassed will be 80%.

Let us now put the data in the entropy function:

$H(Y) = -2 \cdot (0.5 \cdot log0.5) = 0.30103$

$H(Y \mid X_1) = -0.6 \cdot (0.8 \cdot log0.8 + 0.4 \cdot log0.8) = 0.142022$

$H(Y \mid X_2) = -0.4 \cdot (0.2 \cdot log0.2 + 0.6 \cdot log0.6) = 0.109161$

$H(Y \mid X) = H(Y \mid X_1) + H(Y \mid X_2) = 0.142022 + 0.109161 = 0.251184$

This means that using the indicator of blushing allowed us to reduce the level of uncertainty regarding one's feeling embarrassed from an uncertainty (entropy) of 0.30103 to an uncertainty of 0.251184 (information gain of 0.049846) and respectively reduce one's privacy in that aspect. In this example, I have used the entropy function to show that the addition of information helps reduce uncertainty, i.e., adding relevant information about an individual decreases their privacy. This example also demonstrates the ability of AI systems to expose parts of Popper's World 2, **changing the nature of claims made about World 2 from subjective to objective in the same sense that claims about World 3 are objective**, meaning that there is a wide consensus regarding their truthfulness (in this case – regarding one's feeling embarrassed).

Thus, the entropy function can be a measurement of the violation (or non-violation) of personal information privacy, as well as a mathematical tool for creating a metric of the privacy level of personal information.

The Influence of the Network Structure on the Decrease in Order in Personal Privacy

Metcalfe's law states that the financial value of a telecommunications network is proportional to the square of the number of connected users of the system (n^2).

According to Carl Shapiro and Hal Varian, the value of a network is the number of connections between its nodes;[488] formally: network value $V = \frac{n \cdot (n-1)}{2}$,[489] n being the number of nodes. Therefore, if the network value for a single user is \$1, a network of 100 users will have a total value of \$4950.[490] According to the law of large numbers, the network value will eventually asymptotically approach n^2. The influence of the increase in network value on privacy is that more and more personal information is being dispersed to the other nodes in the network, so based on Metcalfe's law, the value of privacy can be said to reduce proportionally to $\frac{1}{n2}$. Thus, the value of personal privacy (pp) for all the network users will formally be $pp = \frac{1}{\frac{n \cdot (n-1)}{2}}$. In a network of 100 users, it amounts to $\frac{1}{4950} = \$0.0002$.

Stuart Kauffman added to Metcalfe's law the parameters k – the number of connections between the nodes, and p[491] – the rules (norms)

---

[488] Carl Shapiro and Hal R. Varian, *Information Rules: A Strategic Guide to the Network Economy* (Boston: Harvard Business school press, 1999), 184.

[489] Based on the combinatoric calculation that if a network has n nodes, they are theoretically connected by $\frac{n \cdot (n-1)}{2}$ edges.

[490] Facebook allows each user a maximum of 5000 friends – a limitation one can bypass by using multiple online identities.

[491] Razi and Yehezkeally, *Ha-olam eino liniari…*, 123.

of connecting by which each node is guided, and which determine the content that will be exchanged between the nodes.[492] I am assuming that $p = 1$, i.e., that private personal information is always exchanged; therefore, the number of connections each network user has with other Internet users by means of ICTs is $k < 1$.[493] This makes the financial value of the network $V = k \cdot \left(\frac{n \cdot (n-1)}{2} \right)$; respectively, $pp = \dfrac{1}{k \cdot \left(\frac{n \cdot (n-1)}{2} \right)}$.

Thus, the very nature of the network social structure formed in the age of ICTs contributes to the ongoing process of depreciation of personal information. This is in line with the earlier presented conclusion regarding the parameters of order and disorder in personal privacy[494] and with the Internet companies' estimate of the financial value of personal information, manifested in the fact that "users' web browsing history elements are routinely being sold off for less than $0.0005".[495]

---

[492] For example, the criteria you use to decide whether to accept or ignore a Facebook friend request, or whether to allow an app to use your location data.

[493] E.g., the number of web applications – such as Instagram, Twitter, Facebook, Google Search, etc. – one is using.

[494] The intensity of information flow between the nodes, the variation in operation modes, the level of interaction between the nodes, the nodes' motivation and involvement level, and the system's hierarchical structure (the Internet network supports a flat organizational structure).

[495] Lukasz Olejnik, Tran Minh-Dung and Claude Castelluccia, "Selling Off Privacy at Auction", 23-26.

The Level of Trust Grows in Direct Proportion to the Amount of Interactions

According to Robert Axelrod, the more two people interact with each other, the greater their mutual trust.[496] The same goes for web applications: the more we use an app, the more we trust it. This explains the indifference with which people provide elements of their personal information to web apps: they believe, based on cumulative personal experience, that the apps are trustworthy and are using the personal information fairly to optimize their services without really violating users' personal privacy. However, this is usually an illusion, as explained in the section about the network market, which exploits our personal information (according to Zuboff – without our knowledge) for profit with little or no consideration for our personal privacy.[497] The conclusion that follows is that the psychology of people's online behavior is another factor that contributes to increasing violation of and disorder in personal privacy.

I would like to propose a metric based on Ohm's law for measuring the current personal information replication online [498]:

Ohm's law states that $I = \dfrac{V}{R}$.

---

[496] Robert Axelrod, *The Evolution of Cooperation* (New York: Basic Books, 1984), 174-175.

[497] Zuboff, *The age of surveillance capitalism*, 178.

[498] https://en.wikipedia.org/wiki/Ohm's_law;

Asher Idan, *Ha-koakh ha-khamishi: Sod ha-reshatot she meshanot et ha-olam* [The Fifth Force: The Force of Networks] (in Hebrew) (Rishon LeZion: Yedioth Ahronoth and Hemed Books, 2014), 159.

I is the current of personal information that is being replicated between the spheres of personal privacy.[499]

V is the information gap that results from the difference in the amounts of personal information between spheres of personal privacy[500] or between one of the spheres of personal privacy and one of the public spheres.[501] Such information gaps are understandable and inevitable[502] because the basic components of personal privacy (body, mind, actions, relationships with external entities, property, and external entities) continuously produce private personal information. Parts of that information, particularly those parts that originate in Popper's World 2 (e.g., thoughts, emotions and desires), are produced at a high rate, and most of them never leave the private sphere. Other parts, which originate in Popper's World 3 (e.g., actions, relationships with external entities, body-related information and property information), are produced at a lower rate. Based on the law of entropy, private personal information is bound to be replicated between the spheres, but the current replication will always be smaller than the current production of private personal information.

R is the resistance to information replication between the private and public spheres; in the context of personal information dissemination through the Internet, it is the resistance to the replication of private personal information from the private sphere to the public sphere. The resistance includes legal barriers (laws and regulations),

---

[499] As defined in Chapter 8.

[500] This book defines the following spheres of personal privacy: Deep personal privacy, general personal privacy, and the little hump.

[501] This book defines the following public spheres: The public sphere, and the big hump.

[502] As shown in the section on the entropy function and uncertainty level.

social barriers (social norms) and psychological barriers (e.g., distrust and fear of shaming), as well as technological barriers specifically designed to protect the privacy of personal information, such as data encryption, blockchain technology, and mathematical principles of big data management like k-anonymity and differential privacy.[503] However, as I have shown in this book, these barriers are incapable of completely preventing the dissemination of private personal information. In the age of ICTs, R is gradually decreasing. There are a few possible ways to calculate and measure R, e.g., a financial metric of how much people would be willing to pay to protect their personal information or the technological tools for data protection (passwords, personal data encryption, storage of personal information on personal computers instead of cloud services, limiting the number of connections to social media, etc.) that people actually use.[504]

Here is a numerical example of the application of Ohm's law for calculating the current personal information replication from the sphere of deep personal privacy to the sphere of general personal privacy:

Let us assume an individual has 10 units of mind-related information (such as feelings, desires, memories etc.) residing in deep personal privacy.

Let us also assume that this individual's confidant (say, their therapist) knows 7 of them.

---

[503] For a further discussion, see Appendix C.

[504] For the purposes of this book: $R = 1$ means one is indifferent to the replication of one's personal information; $R > 1$ means one shows some resistance to the replication of one's personal information; $R < 1$ means one actively contributes to the replication of one's personal information.

That makes the information gap $V = 10 - 7 = 3$. If we substitute it in the entropy function, we will get: $1 - H(v) = 0.52109$

If $R = 1$, then $I = 0.52109$

If $R > 1$, e.g., $R = 1.5$, then $I = \dfrac{0.52109}{1.5} = 0.347393$

If $R < 1$, e.g., $R = 0.5$, then $I = \dfrac{0.52109}{0.5} = 1.04218$

As we can see, Ohm's law is an adequate metric for calculating the current replication of privacy information.

To sum it up, personal privacy is a complex system that is currently in a high degree of disorder due to the increase in entropy, which is accelerated by ICTs and the structure of the network society. In this chapter, we have seen how the entropy function can be used to create a metric of the degree of disorder in personal privacy and how Ohm's law can be used to measure the process of the decrease in order.

Chapter 11

Toward a New Paradigm of Personal Privacy

According to Thomas Kuhn, the replacement of an existing paradigm with a new one, much like a political revolution, is inaugurated by a growing sense that the existing paradigm has ceased to adequately meet the problems that are at its core.[505] In the age of ICTs, we are witnessing a social and economic revolution that leans on a technological revolution where information is the primary resource. The general crisis resulting from this revolutionary state has also touched the paradigm of personal privacy, as we can see from the endless debates and contradictory statements regarding the need for personal privacy and people's attitudes toward their personal privacy. I will begin by describing the cracks in the current paradigm of personal privacy and proceed to outline the components of a new paradigm that may be able to manage the crisis.

---

[505] Kuhn, *The Structure of Scientific Revolutions*, 92.

The Cracks in the Current Paradigm of Personal Privacy

To identify the cracks in the current paradigm, let us review its components (described in detail in Chapter 7) and discuss the validity and problems of each in the age of ICTs.

Symbolic generalizations: An overview of the symbolic generalizations – the theoretical basis of the existing paradigm – reveals that they already contain the "seeds of destruction" of the very paradigm they are part of. Here are some examples:[506]

1. Moore's law, Gilder's law (according to which the total bandwidth of communication systems triples every twelve months) and Metcalfe's law (which defines the financial value of a network) are laws (theories) that predict the exponential growth of the network, the increase in its financial value, and the consequent increase in the flow of personal information from the private sphere to the public sphere.[507]

---

[506] Ibid., 182-183: "One important sort of component I shall label 'symbolic generalizations', having in mind those expressions, deployed without question or dissent by group members, which can readily be cast in a logical form [...] Sometimes they are found already in symbolic form: $f = ma$ or $I = V/R$. Others are ordinarily expressed in words: 'elements combine in constant proportion by weight', or 'action equals reaction'."

[507] Idan, *Ha-koakh ha-khamishi...*, 188.

2. The law of entropy – expressed through the entropy function – mandates the dispersal of personal information, which leads to violations of personal privacy.[508]

3. Big data, organized in big data pools, physically stored in server farms all over the world by means of cloud architecture. Pieces of personal data, thus disseminated among the server farms, are out of the control of the data subjects. ICTs make it possible to fuse all the cloud-stored data in violation of personal information privacy requirements.

4. Profiling: Using people's personal data as an economic resource for customized, targeted advertising, invading people's private space.

5. The IoT: A technology that allows nonstop monitoring of individuals, violating their privacy and making use of both direct and indirect personal data.

6. *K*-anonymity and differential privacy: Formal models designed to protect the anonymity of users whose data are stored in a database.[509]

7. Deep learning: A discipline in computer science that seeks to simulate the decision-making power of the human brain.[510] Deep learning techniques enable the exposure of deep personal privacy.

8. Data encryption: A range of mathematical tools and methods designed to keep information encrypted for

---

[508] Shannon, "A Mathematical Theory of Communication".

[509] Sweeney, "*k*-Anonymity: A Model for Protecting Privacy", 557-570. For a more detailed explanation of these models, see Appendix C.

[510] https://www.ibm.com/topics/deep-learning

anyone who is not authorized to know it in order to protect individuals' privacy online and in big data pools. However, privacy-violating decryption techniques have already been developed as part of the technological "arms race."

9. Social distance: The unseen radius around a person, which marks the social distance necessary for maintaining proper social relations. I have categorized social distance as a symbolic generalization because, in Chapter 2, I propose a metric for its calculation (in which I replace the term "social distance" with "extent of non-knowledge of personal information"), which can be used to measure the extent of an individual's privacy.

As mentioned, the symbolic generalizations of the current paradigm of personal privacy are themselves challenging the paradigm they are part of.

The metaphysical parts of the paradigm of personal privacy are being challenged by the four fundamental social transformations that are the fruit of the digital age:

A. Shifting from the primacy of entities over interactions to the primacy of interactions over entities;

B. Blurring the distinctions between human, machine and nature;

C. Reversal from information scarcity to information abundance;

D. Blurring of the distinction between reality and virtuality.

These social transformations are challenging the metaphysics of the current paradigm of personal privacy in the following ways:

1. Each individual human has a unique selfhood. In the age of ICTs, the concept of unique autonomous selfhood is being challenged for a number of reasons:

- Transformation A undermines individuals' control of their autonomy as the "center of mass" shifts from individuals to interactions.

- The blurring of the distinction between reality and virtuality undermines the uniqueness of our selfhood: now, our selfhood seems to reside not only in our unique individual body; parts of it (like memories, medical data, etc.) are stored on virtual machines and dispersed through the network, where they can be replicated again and again, raising the question whether a selfhood that is dispersed all over the network and can be copied and replicated is still unique and individual. Moreover, the ability to have multiple identities, particularly online identities that reside in the Internet network, outside ourselves, and can be copied and replicated, raises doubts as to the uniqueness of our identity. Lastly, when an online identity can be associated with more than one individual, the connection between selfhood and identity is lost. All these challenge the concept of unique selfhood further and further.

- In a reality where pieces of our personal information are dispersed between multiple nodes in the network, controlling our own personal information becomes virtually impossible.

2. The modern conception of personal privacy is the product of the liberalistic principles of liberty, equality and justice for all. However, when individuals are perceived as nodes in the network, the distinction between private and public is blurred.

3. The concept of identity, which is supposed to be a derivative of our unique individuality and of our right to fulfill it, is becoming unclear for several reasons:[511]

    - The very definition of an individual, their boundaries and their natural rights is becoming vague. For example, what is the status of a cyborg whose boundaries exceed the human body? And what are the boundaries of the human body?

    - The ability to have multiple online and offline identities, whether simultaneously or linearly, is undermining the conception of individual human identity and one's right to fulfill it and disrupting the balance between personal privacy and the public right to know who is hiding behind the online identity – for example, when there is a risk that a certain identity may endanger public safety.

4. Private and public spheres: the private sphere consists of the physical and virtual areas where one has the right to fulfill one's self-potential, while in the public sphere, one is bound by the rules and norms of society:

    - The distinction between the private and public spheres is becoming blurred in terms of space and

---

[511] Rabi, *Omes ha-individualiyut*, 255.

time division. The two most prominent manifestations of this are work from home and prosumption. The blurring of the distinction between the private and public spheres reduces our ability to maintain and realize our autonomy in our private physical space.

- Pieces of our personal (and private personal) information are dispersed throughout the network, out of our private sphere and out of our control.

- The reversal from information scarcity to information abundance due to ICTs that allow the free flow of information[512] leads to private personal information flowing out of the private sphere, where it was created, into the public sphere.

- The network structure of ICTs supports the strong flow of personal data from the private sphere to the public sphere.

- The status and ownership of data aggregated in big data pools is in debate: should they be owned and controlled by the data subjects or by the data pool owners?

5. The individual as an agent: The blurring of the distinctions discussed in the previous paragraphs undermines our agency, both in terms of individuality and in terms of social norms.

---

[512] This idea is in line with the meme theory, which regards human beings as vessels for memes, whose primary purpose is to spread information and ideas.

6. The ICT revolution is spreading quickly, violating the privacy of personal information in a variety of ways. Despite the efforts, the attempts to regulate have failed to cope with the effects of this revolution on personal privacy. **The biggest crack in the current paradigm concerning regulation is the failed assumption** that personal information privacy can be protected by regulations and social norms that are based on consent and control. This assumption is failing for the following reasons:

- AI-based algorithms profile users based on their gender, race and other categories, violating the principle of equal treatment. They are basically constructing and analyzing statistical virtual identities that do not necessarily match people's actual identities, taking control over one's identity out of one's hands and labeling people without their consent. This entire process is done without the users' consent and is beyond their control.

- IoT devices used in "smart cities" and processors built into more and more everyday appliances such as refrigerators, air conditioners and cars usually lack a human-machine interface through which one could express consent or non-consent to the collection of personal information. As a result, people's information is collected without their expressed consent and without any ability to control which information will be collected or how and by whom it will be used.

- The various databases in which personal information is stored are interconnected. People do not provide consent to that – whether expressed or implied nor can they control how and where their personal data will be stored or how it will be processed and disseminated. The purposes for which our personal information is used are beyond our control, too, and we are not asked to provide consent for any such use.

- Devices that directly interface with our body a priori bypass the conscious mechanisms of decision-making, so there can be neither control over the personal data not consent to provide it.

- The scope of information available in the age of ICTs is too vast for people to make informed choices about which of all their data they would like to control and which they don't mind being collected. According to McDonald and Cranor's study, people do not read privacy policies, and if they did, the annual cost to the US market would be $781 billion.[513]

7. Social distance: According to Georg Simmel, social distance is the extent of familiarity or unfamiliarity between people.[514] In the age of ICTs, social distance is getting smaller and smaller.

---

[513] Aleecia McDonald and Lorrie Faith Cranor, "The Cost of Reading Privacy Policies", *A Journal of Law and Policy for the Information Society* 4, no.3 (2008): 543-568.

[514] Simmel and Wolff, *The Sociology of Georg Simmel*, 312.

8. Personal information: ICTs and the network structure of our society have made personal and private information much more available and accessible to others.[515]

9. The current paradigm of privacy does not acknowledge that personal privacy is a complex system with a defined order. ICTs have caused this system to be in a high degree of disorder, which the current paradigm is incapable of managing.

Privacy regulation that relies on consent and control has proven impractical and unenforceable.[516] The reason for its failure is the overabundance of information. People do not know all the information that is being collected about them – nor do they want to know. They do not understand all the consequences of information collection – nor do they want to. They do not have a real choice – or the will to choose – regarding the information that will be collected. As a result, people are making wrong decisions regarding the privacy of their personal information and selling it cheap – or giving it out for free – to Internet companies. People have no actual control over their personal information. The mechanisms of consent and control are being protected, but the actual information privacy is not.[517]

---

[515] In Chapter 8, I have shown how ICTs are turning private information from a metaphysical entity into a symbolic generalization in the emerging new paradigm.

[516] Lalana Kagal and Hal Abelson, "Access Control Is an Inadequate Framework for Privacy Protection", MIT Computer Science and Artificial Intelligence Lab, 2010.

[517] Government and regulatory agencies impose severely fines on Internet companies, but as I have previously shown, it does not help protect information privacy.

In Chapter 7, I have described the values on which the current paradigm of personal privacy is based. Let us now review these values and see how they are challenged in the age of ICTs:

1. Universal right to self-fulfillment: This right is challenged in the network society, where every individual is a node in the network, and there is no distinction between the private and public spheres.

2. The liberalistic principles of liberty, equality and justice for all are becoming harder to realize in view of the increased monitoring of individuals by governments and Internet companies (as proven by Snowden's documents and the Facebook-Cambridge Analytica data scandal).

3. The duties of individuals and society: In this age of multiple identities, individual rights are often considered to include the right to have multiple identities, even though some of those identities may contradict the public interest.

4. The balance between the individual and society is disturbed by the advent of surveillance capitalism, which regards personal information as an economic resource free for the taking that may be used for profit without any supervision.

5. The state's role as "the watchman" that only interferes where the individuals fail to settle matters between themselves is being questioned as states exploit ICTs and terrorism threats to turn themselves into a "big brother" that invades people's private sphere with hardly any restrictions.

6. Individual anonymity is practically nullified in the age of ICTs because of non-stop monitoring by IoT devices, big data and network connectivity. The concept of a "smart

city" includes the deployment of numerous sensors (e.g., cameras and computerized traffic light systems) in public areas (mainly urban areas) and connecting them to special apps designed to improve municipal services (e.g., public transportation, water supply and garbage collection). To do that, the apps collect pieces of personal information about every person in a given urban area. That personal information is aggregated in big data pools and analyzed using AI technologies.

Smart cities are violating the conception of consent and control in a number of ways: for one thing, smart city data collection systems do not have an expressed privacy protection policy, even though most web apps are required to have one and to have the user consent to it as a necessary condition for installation and use. Furthermore, these systems usually have weak protection against cyber-attacks. Thus, smart cities aggregate and use residents' personal information without their prior consent, analyze it and share it with numerous apps without limiting its use to the purposes for which it was collected and without having proper protection as required by data security regulations.

Those are the ways in which ICTs are challenging the values of the current paradigm of personal privacy.

In Chapter 7, I described a few exemplars related to personal privacy.[518] Within the current paradigm of personal privacy, these are seen as mere challenges that the paradigm can overcome; however, the

---

[518] A short list of prominent case studies or models of problem-solving.

following (and other) examples indicate its inability to effectively handle those challenges:

1. The leaking of the personal data of 50 million Facebook users by Cambridge Analytica Ltd. to be unfairly used in Donald Trump's 2016 presidential campaign.[519] This is an example of attempted mind engineering in violation of the unwritten rules of democratic play.

2. In Israel and many other countries, the time of the COVID-19 pandemic was marked by numerous invasions of personal privacy.[520] The Shin Bet used its location app to monitor citizens 24/7 under the excuse of public health protection. This is an example of disruption of the balance between personal privacy (individual rights and liberties) and the state's power over its citizens.

3. The COVID-19 vaccine supply contract between the Israeli government and Pfizer drew the public's attention to the vast amounts of information the Israeli Ministry of Health has about the country's residents – sensitive information that includes every health condition and every treatment ever received.

---

[519] Nicholas Confessore, "Cambridge Analytica and Facebook: The Scandal and the Fallout So Far", *The New York Times*, 4 April 2018 https://www.nytimes.com/2018/04/04/us/politics/cambridge-analytica-scandal-fallout.html (Accessed 5/4/2024)
As a result of the scandal, Facebook founder and owner Mark Zuckerberg was called to testify before the US Congress to provide explanations and present a plan of action for improving Facebook users' privacy protection.
[520] Shakhaf, "2020 hayta shana shkhora la-pratiyut be-Israel".

These are but a few examples that illustrate the crisis in the current paradigm of personal privacy and show how it is failing to cope with the challenges posed by ICTs. The reason for the crisis is the fact that the advent of ICTs has cracked all the main conceptions underlying the current paradigm – Westin's conception of privacy as control,[521] Nissenbaum's conception of contextual privacy,[522] and Gavison's conception of limited access[523] – in all its aspects (symbolic generalizations, metaphysics and values). These cracks cannot be mended using the tools of the current paradigm; they run too deep because they are a product of an all-encompassing revolution started by ICTs.[524] The most important characteristic of this revolution is the transformation of personal information into the primary resource of a new economy. That makes the dissemination of personal information from the private sphere into the public sphere a financial interest, and therefore, the conception and discourse of personal information – and, consequently, of personal privacy – must also transform from discourse of individual rights to discourse of ownership and property. This is the greatest challenge to the current paradigm of personal privacy, which had been founded on liberalistic values centered on individualism and individual rights.

---

[521] Westin, "The origins of modern claims to privacy", 56-59.
[522] Nissenbaum, *Privacy in Context...*, 2.
[523] Gavison, "Privacy and the Limits of Law".
[524] Toffler, *The Third Wave*, 18.

Proposed New Components for a New Paradigm of Personal Privacy

The crisis in the field of personal privacy indicates that we need a new paradigm – a shift in our worldview, a new look at reality.[525] Description of scientific paradigms is usually the work of historians of science, who investigate the common framework of a given discipline at a given time,[526] so it is possible that the new paradigm of personal privacy will receive its final formulation only in retrospect. It is still too early to say whether any new paradigm of personal privacy is already in place, let alone to describe it; those are open research questions for the future. All I can do now is draw an outline for a potential new paradigm of personal privacy that could address the crisis the old one is found in. I believe that this new paradigm will be an extension and continuation of the current one,[527] i.e., instead of contradicting it, it will provide new meaning to some familiar concepts.[528]

The following outline is a prediction based on the current state of personal privacy, considering the known and expected developments in the field of ICTs, as described in this book.

Symbolic generalizations: The symbolic generalizations[529] of the current paradigm of privacy will probably remain valid, with the following additions:

---

[525] Kuhn, *The Structure of Scientific Revolutions*, 111.

[526] Ibid., 43.

[527] Ibid., 95-99.

[528] Ibid., 102.

[529] Ibid., 182-183: "One important sort of component I shall label 'symbolic generalizations', having in mind those expressions, deployed without question or dissent by group members, which can readily be cast

1. Game theory as a source of economic and social models of balancing between the personal privacy needs of the individual human agent in the network society and the financial and social interests of other network players (Internet companies, corporations and governments). The desired balance will be achieved through equilibrium: Nash equilibrium for competitive games and other forms of equilibrium for cooperative games.[530]

2. Use of chaos theory to describe personal privacy as a complex system in a high degree of disorder that cannot be brought to an equilibrium using game theory models, but is swinging around an attractor.

3. New mathematical models for protecting privacy, particularly for protecting anonymity.

4. Quantitative models for measuring the extent of personal privacy in a given situation.

5. Reintroduction of human supervision and decision-making to automated AI-based processes to prevent racial, gender and other forms of bias that seem to be built into AI algorithms.[531]

---

in a logical form [...] Sometimes they are found already in symbolic form: $f = ma$ or $I = V/R$. Others are ordinarily expressed in words: 'elements combine in constant proportion by weight', or 'action equals reaction'."

[530] For formal game theory definitions, see Appendices A and B.

[531] EU GDPR, Article 22, "Automated individual decision-making, including profiling".
EU GDPR, Recital 71, "Profiling".

6. Privacy protection laws built into AI systems to control the collected data – an extension of the concept of data protection by design or by default.[532]

7. An economic model of Internet company taxation based on the amount of personal privacy information the company uses – similar to electricity and water payments. The model will measure privacy information in kilobytes of information from each category of privacy information – for example, how many kilobytes of medical or financial information Facebook has on a particular individual user or how many kilobytes of medical information in total it has about all its users.[533]

Metaphysical parts: The metaphysical foundation of the new paradigm of personal privacy will be the four social transformations of the age of ICTs. For a more detailed explanation of how the metaphysical parts are built on them, see Chapters 7-9.

1. The information society: The network society is a product of the Third Wave revolution.[534] Its primary metaphysical components are the network economy – composed of network production, network work and the network market; surveillance capitalism, which exploits personal information to maximize financial profit; and the

---

[532] Because of legal restrictions imposed on AI systems by Western countries (e.g., the GDPR), there is a fear that Western AI technologies will fall behind AI technologies built by countries that have less consideration for privacy protection, such as China.

[533] The model may be progressive, like electricity and water consumption billing.

[534] Toffler, *The Third Wave*, 18.

economics of privacy – a subset of the information economy focused on personal privacy as an economic resource.[535]

2.   The network human: This will be the main change in the new paradigm of personal privacy. The concept of network humans will replace the concept of autonomous, unique individuals with rights that include the right to personal privacy.

3.   Personal privacy as defined by this book: Personal privacy divided into general privacy and deep privacy, with a respective distinction between personal information and private personal information; personal privacy made up of several basic components that are related to each other and together cover all the aspects of privacy; personal privacy as information, with a consequent shift from discourse of personal privacy to discourse of personal information and private personal information; and personal privacy as a complex system in a high degree of disorder, to manage which we will need to find new forms of equilibria, or at least identify attractors around which the system might settle.

4.   Privacy information is composed of memes, which have the characteristics of a life form, and therefore, it is bound to be dispersed throughout the network subject to the law of entropy and the laws of the network.

5.   A new regulation system based on the reasonable and proper use of personal information rather than on the

---

[535] Acquisti, Taylor and Wagman, "The Economics of Privacy".

explicit consent of the data subject, in accordance with the following principles of data collection and use: the notification and consent mechanism will only apply in extreme cases (be an exception rather than the rule) and only when the information has significance; because in the age of ICTs information and data always have additional secondary users who may be affected by the personal data in unpredictable ways, risk assessment and analysis must include the risks that stem from the data handling and processing rather than from the raw data itself, and be performed by the data users (the Internet companies) and not by the data subjects. Shifting the focus from the information content to its significance and requiring the data subject's consent reduces the need to prove that the elements of personal information that are being collected are exactly the ones that are needed for the declared purposes. Highlighting the uses made of the personal information (which is more or less the same as the purposes of collection), including undeclared secondary uses, increases the data users' accountability and commitment to use the data fairly without causing harm to the data subjects. **The new regulation system should do with protecting deep personal privacy. Once we acknowledge the fact that personal information is a vital resource in the Third Wave economy (which includes the network economy), we should develop mechanisms to ensure fair trade of this resource.**

These principles are shifting the focus from consent and control by the data subject to the ways Internet companies use personal information. In other words, the question of whether to allow the use of the information passes from the data subject to the data user.

Values: The new paradigm of personal privacy will not necessarily have new values; therefore, the purpose of this section is to outline the potential development of the current values of personal privacy in line with the new conception of personal privacy that is taking form in our time.

1. Universal right to self-fulfillment: In a network society, where every individual is a node in the network, and there is no distinction between private and public spheres, no basic component of personal privacy remains exclusively the individual's private business anymore. Therefore, individuals have the right to fulfill themselves as long as it does not compromise their function as network nodes.

2. The liberalistic principles of liberty, equality and justice for all – the basic guidelines of individual and social behavior in a liberal democratic society – will change from absolute values into relative principles that acknowledge the importance of individual rights on the one hand, and "the butterfly effect" of Internet hyperconnectivity which allows viral spread of fake news, terrorist and racist movements, and the like on the other hand, balancing between increased supervision of online content and the risk of evolving into a "big brother."

3. The acknowledgment of deep privacy as an existential human need and the creation of an ethical and legislative environment that would ensure its protection.

4. The duties of individual and society: Increasing legitimacy of multiple identities as beneficial for the individual and for society alike.

5. Redefined balance between individual and society: the acknowledgment of the fact that personal information is a financial resource that may be used for profit and ensuring fair trade of that resource with the information subject.

6. The acknowledgment that anonymity in the public sphere is limited to secret elections and prohibition of targeted political brainwashing based on personal characteristics exposed by means of ICTs (as Cambridge Analytica had done)

7. Ethics and legislation that prohibit the exposure of deep personal privacy beyond the confidants' strict need to know.[536] Deep privacy protection practices will be enforced – and coerced, if needed – by government agencies based on the recognition that some personal information, like genetic information, must remain confidential for all our sakes.

Exemplars: Previously, I described several examples of the crisis in the current paradigm of personal privacy. Since there is no actual new paradigm of personal privacy yet, there are no characteristic exemplars or models of problem-solving to list. I can only predict that those will be along the following lines:

---

[536] Anita L. Allen, *Unpopular Privacy: What Must We Hide* (New York: Oxford University press, 2011), Abstract.

1. Government attempts to place restrictions on large Internet companies in order to balance their corporate interests against the public interest:

- In 2021, Facebook founder and CEO Mark Zuckerberg and then CEO of Twitter Jack Dorsey appeared before the US Senate (for Zuckerberg, it was not the first time) to defend the famous social network's speech moderation actions during the 2020 election.[537]

- In 2019, Google was fined €1.5 billion by the EU – its third fine since 2017 – for antitrust violations in the online advertising market.[538]

2. Attempts to identify and combat the spread of fake news are already being made to prevent unfair bias from being created and spread throughout the network by means of social media.[539]

Like the previous examples, these indicate the probable direction in which the new paradigm of personal privacy will develop.

---

[537] "Zuckerberg and Dorsey Face Harsh Questioning From Lawmakers", *New York Times*, 17 November 2020. https://www.nytimes.com/live/2020/11/17/technology/twitter-facebook-hearings (Accessed 27/5/2024)

[538] Adam Satariano, "Google Fined $1.7 Billion by E.U. for Unfair Advertising Rules", *The New York Times*, 20 March 2019. https://www.nytimes.com/2019/03/20/business/google-fine-advertising.html (Accessed 27/5/2024)

[539] Hunt Allcott and Matthew Gentzkow, "Social Media and Fake News in the 2016 Election", *Journal of Economic Perspectives* 31, no. 2 (Spring 2017): 211-236.

I have now given you my general prediction regarding the new paradigm of personal privacy, which would be more compatible with the characteristics of the age of ICTs. This new paradigm should have to do with protecting deep personal privacy, which is a fundamental human need; any privacy information beyond that cannot be protected due to the network structure of society and the nature and power of ICTs. Therefore, instead of trying to protect general personal privacy, we must seek to establish fairer relations between individual and society and between individual and other entities in order to find a new balance between their different needs.

Chapter 12

Summary and Conclusions

The topic of this book is personal privacy in the age of ICTs. For the last two decades, it seems to be in the focus of public attention constantly. Every now and then, we hear of other violations of cyber privacy by companies, well-known organizations, and even governments.[540] Hackers breach people's privacy for personal gain or as others' proxies; commercial organizations (Internet companies) do it for profit, using personal information as an economic resource; governments seek to increase surveillance over individuals (citizens and foreigners) for political ends, using protection against terrorism or other threats (such as COVID-19) as excuses. Due to these repeated violations of personal privacy, some argue that it is not that crucial to people, bringing as evidence the ease with which people seem to waive their personal privacy in exchange for services provided by large Internet companies. On the other hand, we are seeing efforts by government and regulatory agencies to protect personal privacy. In the US, Mark Zuckerberg was called before the Senate to explain Facebook's privacy policy, while the EU passed the GDPR in an attempt to protect Europeans' personal privacy. Either way, the issue of privacy, particularly various aspects of personal privacy, persistently remains on the public agenda.

In this book, I tried to clarify some basic concepts and answer some fundamental questions regarding personal privacy: What is

---

[540] Prominent examples are mentioned throughout this book.

personal privacy? Where and to what does it apply? Where does the private sphere end and the public sphere begin? Where does privacy stand in relation to other concepts, such as freedom, autonomy and liberty? How to balance individual rights against the public's right to know? Must a state reveal potential threats to its security? And how should it all be regulated in this age of information and communication technologies that follow us wherever we go and whatever we do? To be able to answer these questions and to discuss the influence of information and communication technologies (ICTs) on personal privacy, I first had to clarify what personal privacy was. I divided the concept of personal privacy into **two categories – general personal privacy and deep personal privacy – to show that throughout human history and across different societies, general personal privacy has always been violated one way or another, sometimes even by law, while deep personal privacy, hidden in our body or mind, had been virtually inaccessible and therefore never violated. The greatest change of our time with regard to personal privacy is that deep personal privacy has also become partially accessible by means of ICTs. To support my arguments, I chose to shift the discussion to deep and general personal privacy information for the following reasons: every basic component of personal privacy can be translated into personal privacy information; information is a much more clear, measurable and quantifiable concept;[541] information and communication technologies – as suggested by their name – are technologies that handle and process information.**

---

[541] In Chapter 2, I introduced ways to quantify and measure personal privacy information.

Up until the age of ICTs, deep personal privacy had been inaccessible to others and had, therefore, never been culture-dependent. This has changed now, with ICTs using personal privacy information to increasingly uncover our deep personal privacy.

The advent of ICTs has created a new civilization Alvin Toffler calls the Third Wave: a society in which knowledge and information are the primary resources and determinants of power, and many of the human activities are done by virtual means using private personal information.[542] That information is being stored, fused, analyzed and shared to support our way of life and serve us as individuals and as a society on the one hand, but on the other hand, to serve the economic and political interests of governments, conglomerates and social networks. The Third Wave society is new and rapidly changing – at a pace that humanity finds hard to follow – and is challenging the existing paradigm of personal privacy, which was shaped during the advent of the Second Wave.

Personal privacy is a fundamental human need that stems from a biological mechanism common to humans and other animals, which is manifested in maintaining distance from other individuals. In humans, the physical distance is translated into social distance, which is a necessary condition for proper social interaction. This means personal privacy is essential for society as much as it is essential for the individual. No society can exist without some degree of personal privacy. There is an apparent line that separates the individual from their membership groups, but that line is flexible, frequently shifting, and time- and place-dependent. Within it resides our personal privacy, and beyond it lies the

---

[542] Online shopping, personal and social communication by social media, sharing of information and knowledge by smartphone, using personal online profiles for political campaigning, and so on.

public sphere. Social distance is a consequence of the information and knowledge people have about each other. **The extent of familiarity or unfamiliarity between people is the sphere of personal privacy**, and it is the very thing that allows social relationships between individual humans.[543] In Chapter 2, I introduced a quantitative metric for calculating the extent of non-knowledge of personal privacy, which allows for measuring the level of an individual's personal privacy in any given situation.

I proceeded by discussing personal privacy in different societies and showing it on the "three-dimensional space" chart – a model that represents a human society as a three-dimensional space whose axes are knowledge, values, and regulations (laws and folkways). I used this model to show how personal privacy is affected by culture and to develop a quantitative metric for intercultural comparison of personal privacy.

In the modern era, the concept of privacy was extended to include the mental as well as the physical sphere. The Western conception of personal privacy is based on the philosophy of individualism – a concept shaped by the greatest philosophers of the Enlightenment (e.g., Kant, Marx, Nietzsche, Spinoza and Rousseau). The philosophy of individualism defines the limits of personal autonomy and the norms of behavior in the private and public spheres. The Enlightenment philosophers seem to have been aware of the seemingly insoluble tension between the individual and society, the unique and the universal. Out of that tension, as a way to resolve it, the concepts of personal privacy and a private sphere were born, in which one is entitled to fully realize one's self-potential. Beyond it lies the public sphere, where one must abide and be bound by the rules and the social norms.

---

[543] Simmel and Wolff, *The Sociology of Georg Simmel*, 312.

This division received legal and normative protection due to the continued awareness of the inherent conflict between personal privacy and public interests (especially matters of public safety).

From the various conceptions and definitions of personal privacy, I extracted the basic components that are present in all of them in varying degrees. Personal privacy is a scale; each point represents a state of personal privacy that is the product of a particular culture and balance of power. The two extreme cases – full privacy and no privacy at all – are impossible. Privacy can also be described as a network in which not everyone can enjoy the same level of personal privacy. Each basic component of personal privacy contains elements of information that together makeup all our personal privacy information. To help me map the division and interaction between the private and public spheres and between privacy and non-privacy, I used Karl Popper's theory of Three Worlds: World 1 – the physical world; World 2 – the mental or psychological world; and World 3 – the realm of the products of human thought.[544] Based on Popper's definitions, I placed the private sphere in World 2 and those parts of Worlds 1 and 3 that are perceived to be private and personal. Then, I defined deep personal privacy as World 2 and general personal privacy as the parts of Worlds 1 and 3 that are considered to be private. Together, all the combined information of both types of personal privacy – deep and general – constitutes personal privacy information. The public sphere begins where the private sphere ends. Everything we do in the public sphere is open and exposed to all. The public sphere is present only in Worlds 1 and 3.

I proceeded by describing the modern liberalistic societies' paradigm of personal privacy, which includes all the various conceptions

---

[544] Popper, "Three Worlds – The Tanner Lecture on Human Values".

and definitions of personal privacy discussed earlier and is based on the conception of human beings as unique autonomous individuals. To see how all the different conceptions and definitions of personal privacy can coherently coexist, we have to understand that they are all part of the same paradigm, which has the basic components of personal privacy and the connections between them at its core.

After that, I began the shift from discussion of personal privacy to discussion of personal privacy information. First, I defined the basic concepts of **deep personal privacy information** as the knowledge one has about oneself minus the knowledge the world (or society) has about one, and **general personal privacy information** as the knowledge one and one's confidants have about one minus the knowledge the world has about one. Based on that, I placed all deep personal privacy information in Popper's World 2 and general personal privacy information in Worlds 1 and 3. To show how different categories of information are related to the basic components of personal privacy, I discussed them in detail, particularly the identity and anonymity components. Some of the basic components of personal privacy have received new meaning in the age of ICTs.

I then described the properties and characteristics of information in general, and personal privacy information in particular, and showed how personal privacy information fits the definition of information when it is part of a general paradigm of personal privacy. Shifting the focus of the discussion to personal privacy information allowed me to better explain the enormous impact of ICTs on personal privacy.

Around the end of the second millennium, a number of major social, technological, economic, and cultural transformations came together to give rise to a new form of society – **the network information**

society, which has its own unique characteristics.[545] The economy of the network society is different from the Fordist economy and based on a paradigm of "flat" decentralized production, with a production and supply network that is often spread across different continents. The decentralization of the production process is associated with a methodological shift from bulk mass production in large standardized batches to mass customization,[546] which is characterized by shorter production cycles, smaller batches and greater adaptability to the market's changing demands.[547] This new network production and service model requires that vendors and customers have intimate knowledge of each other.[548]

Network production consists of two layers of networks: the first layer is a physical network comprised of nodes that exchange physical goods, while the second layer is an information network in which the nodes exchange information. Some of the prosumer information that flows through the network is private. Customized production and services require the producer to have intimate knowledge of the consumer. Network production breaks the production-consumption and producer-consumer dichotomy. A node in the production network can be in the public sphere and in the private sphere interchangeably or even simultaneously without changing its physical location. The private-public dichotomy that existed in the industrial society is challenged by the growing blur between workspace and private space, work time and

---

[545] Castells, *The Rise of the Network Society*, xvii.

[546] Dollarhide, Mass Customization: "Mass customization is the process of delivering market goods and services that are modified to satisfy a specific customer's needs."

[547] Robins and Webster, *Times of Technoculture*, 150-151.

[548] Fisher, *Capitalism be-idan ha-tikshoret ha-digitalit*, 127.

leisure time, and the flow of private personal information between the network nodes.

Network work has three main characteristics: a blur of the distinction between work and non-work, workspace and living space, work time and personal time; a blur of the distinction between managers (capitalists) and workers; and the growing importance of skill and professionalism and of network technologies. Because the network workers' time and location are flexible and not necessarily bound to the production process, work can invade their private time. Thus, network technologies are blurring the distinction between work and leisure, productive time and free time, office space and home space.

The network market is creating a new kind of economy – a network economy based on communication, where the Internet is the dominant technology that allows every individual to become a node in the economic network and to communicate with other nodes in order to perform its financial operations. Every individual node is "dumb," i.e., unable to see and comprehend the network in its entirety. The new economy uses network technology to remove the obstacles (norms and regulations) between the individual and the market. Its primary resource and fuel is our personal information.

Large Internet companies exploit our personal information to maximize their profits. They use it to develop targeted advertising and, eventually, niche production and niche marketing. Their means of production are AI technologies, which help them improve their products and services by analyzing big data that reflects customer preferences. This is how our personal information is turned into a resource that translates into a bottom line. In order for it to be used and traded as described, it must be taken out of the private sphere and out of our control and pass on to other players in the market. The market, by definition, resides in the public sphere (except in specific settings in which the information is exchanged between confidants), and so the public sphere invades the private sphere to serve the market's financial interests.

The economics of privacy is a subset of the information economy that focuses on the economic changes resulting from the protection or disclosure of private personal information. The widespread use of web applications allows the collection of personal information that before the Internet was inaccessible and remained in the private sphere. Disclosure or non-disclosure of your personal information has both positive and negative financial implications – not only for you but also for the entities you are dealing with.[549] This raises a number of questions: Can the benefits of personal information disclosure for the information subject and the other parties to the information exchange be balanced against its drawbacks? How do Internet companies profit from our personal information? What would be the right balance between the economic value of personal information that is being aggregated in big data pools and traded in conditions of a free market that regards privacy as an economic commodity and legislation and social norms that regard privacy as a fundamental human right? In Appendices A and B, I try to answer these questions using economic and game theory models.

The advent of ICTs has had an enormous impact on every aspect of our lives. The age of ICTs has created a new culture and new human experiences that necessarily change the way we perceive our body and our identity and redefine our self-conception (who we are), our mutual interactions (how we socialize), our conception of reality (our metaphysics), and our interactions with and impact on reality (our

---

[549] See Appendices A and B for the game theory models I am proposing to use to find balance between the pros and cons of disclosure and non-disclosure of personal information.

agency). **Each of these four basic components of our worldview affects our conception of personal privacy and how we live it out in the world.**

Our personal information (personal privacy information) is bound to leak out to other entities (be it other individuals or any group or organization) for several reasons:

According to Richard Dawkins' meme theory, information – including personal privacy information – "wants" to disperse and replicate itself. This property of information is one of the reasons why private personal information tends to leak out of the private sphere into the public sphere in violation of personal privacy. In Appendix A, you can find game theory models that illustrate the rationale of personal information replication and dissemination among the players in the information network.

Personal privacy can be explained as a complex system that has a certain order and its violations as disruptions of that order, i.e., manifestations of growing entropy. The age of ICTs has personal privacy in a state of turbulence, which is not a point on the spectrum between order and chaos but a whole different range of states characterized by "whirlpools" that disrupt the existing order of things until a new order is formed. I described the five parameters for measuring the degree of order or disorder in a complex system, such as personal privacy, and used them to show that the system of personal privacy is in a very high degree of disorder, which creates an intense feeling of chaos in this field.

The advent of ICTs has cracked all the main conceptions underlying the current paradigm of personal privacy – Westin's conception of privacy as control,[550] Nissenbaum's conception of

---

[550] Westin, "The origins of modern claims to privacy", 56-59.

contextual privacy,[551] and Gavison's conception of limited access[552] – in all its aspects (symbolic generalizations, metaphysics and values). These cracks cannot be mended using the tools of the current paradigm; they run too deep because they are a product of an all-encompassing revolution started by ICTs.[553] The dissemination of personal information from the private sphere into the public sphere has become a financial interest, and therefore, the conception and discourse of personal privacy must also transform **from the discourse of individual rights to the discourse of ownership and property**. This is the greatest challenge to the current paradigm of personal privacy, which had been founded on liberalistic values centered on individualism and individual rights.

In Appendix C, I list the mathematical and technological obstacles that prevent the mending of the cracks in the current paradigm of personal privacy. This suggests that we need a new paradigm, one more compatible with the characteristics of the age of ICTs. An effort to shape a new paradigm should include a redefinition of its content. In my opinion, the new paradigm should be aimed only at protecting deep personal privacy, which is a fundamental human need, because the nature of the network society and ICTs makes the protection of general personal privacy impossible.

---

[551] Nissenbaum, *Privacy in Context...*, 2.
[552] Gavison, "Privacy and the Limits of Law".
[553] Toffler, *The Third Wave*, 18.

References

Books

Allen, Anita L. *Unpopular Privacy: What Must We Hide? (Studies in Feminist Philosophy)*. 1st edition. New York: Oxford University Press, 2011.

Andrejevic, Mark. *Infoglut: How Too Much Information is Changing the Way We Think and Know*. New York: Routledge, 2013.

Axelrod, Robert. *The Evolution of Cooperation*. Revised edition. New York: Basic Books, 2006.

Bateson, Gregory. *Steps to an Ecology of Mind*. New York: Ballantine Books, 1972.

Bell, Daniel. *The Coming of Postindustrial Society: A Venture in Social Forecasting*. New York: Basic Books, 1973.

Benkler, Yochai. *The Wealth of Networks: How Social Production Transforms Markets and Freedom*. New Haven and London: Yale University Press, 2006.

Birnhack, Michael D. *Merkhav Prati: Ha-zkhut la-pratiyut ha-ishit beyn mishpat le-technologia* [Private Space: The Right to Privacy, Law and Technology] (in Hebrew). Ramat Gan: Bar-Ilan University, 2010.

Blackmore, Susan. *The Meme Machine*. Oxford: Oxford University Press, 1999.

Castells, Manuel. *The Rise of the Network Society*. Oxford: Wiley-Blackwell, 1996.

Dawkins, Richard. *The Selfish Gene*. Oxford: Oxford University Press, 1976.

Dewey, John. "Ethics" (1908), in: *The Middle Works of John Dewey*, ed. Jo Ann Boydton. Carbondale: Southern Illinois University Press, 1978.

Emerson, Ralph Waldo. "Fortune of the Republic" (1878), in: *The Works of Ralph Waldo Emerson in 12 vols, Fireside Edition, Vol. 11 ("Miscellanies")*. Boston and New York, 1909.

------------------------------"Politics" (1844), in: *The Works of Ralph Waldo Emerson in 12 vols, Fireside Edition, Vol. 3 ("Essays: Second Series")*. Boston and New York, 1909.

Etzioni, Amitai. *The Limits of Privacy*. New York: Basic Books, 1999.

Fried, Charles. *An Anatomy of Values: Problems of Personal and Social Choice*. Cambridge: Harvard University Press, 1970.

Garfinkel, Simon. *Database Nation: The Death of Privacy in the 21st Century*. 1st ed. Sebastopol, California: O'Reilly, 2000.

Gourley, David and Brian Totty. *HTTP: The Definitive Guide*. Sebastopol, California: O'Reilly, 2002.

Harpaz, Amos. *Sheker Ha-Individualism: Spinoza, Hegel, yeha-dimui ha-kozev shel ha-'ani' ha-moderni* [Falsity of Individualism: Spinoza, Hegel, and the False Image of Modern Man] (in Hebrew). Tel Aviv: Resling, 2013.

Hixson, Richard F. *Privacy in a Public Society: Human Rights in Conflict*. Oxford: Oxford University Press, 1987.

Holmberg, A. R. *The Siriono* (unpublished Ph. D. dissertation). Yale, 1946.

Hongladarom, Soraj. *A Buddhist Theory of Privacy*. Singapore: Springer, 2016.

Howard, H. Eliot. *Territory in Bird Life*. New York: E. P. Dutton and
Co., 1920.

Howells, William W. *The Heathens: Primitive Man and His Religions*.
New York: Doubleday, 1948.

Hume, David. *A Treatise of Human Nature*. Oxford: Clarendon Press,
1896.

Jones, Livingston F. *A Study of the Thlingets of Alaska*. New York:
Kessinger Publishing, 2004.

Kant, Immanuel. "An answer to the question: What is enlightenment?"
(1784), in: *The Cambridge Edition of the Works of Immanuel
Kant: Practical Philosophy*. Cambridge, New York:
Cambridge University Press, 1999.

----------------------*Groundwork of the Metaphysics of Morals* (1785).
Cambridge: Cambridge University Press, 1998.

Karmin, Anatoly. *Culturology: Textbook*. Saint-Petersburg: Lan, 2008.

Keynes, John Maynard. *The General Theory of Employment, Interest
and Money*. London: Palgrave Macmillan, 1936.

Kuhn, Thomas. *The Structure of Scientific Revolutions*. 2nd edition.
Chicago: University of Chicago Press, 1970.

Laurie, Graeme. *Genetic Privacy: A Challenge to Medico-Legal Norm*.
Cambridge: Cambridge University Press, 2002.

Lee, Dorothy. *Freedom and Culture*. Englewood Cliffs, N. J.:
Prentice-Hall, 1959.

Locke, John. *Second Treatise of Government*. London: Awnsham
Churchill, 1689.

Maschler, Michael, Eilon Solan and Shmuel Zamir. *Game Theory*. New
York: Cambridge University Press, 2013.

McLuhan, Marshall. *Understanding Media: The extensions of man*.
London and New York: McGraw-Hill, 1964.

Mead, Margaret. *Coming of Age in Samoa: A Psychological Study of Primitive Youth for Western Civilization*. New York: New American Library, 1949.

Mill, John Stuart. *On Liberty*. London: J. W. Parker and Son, 1859.

Moore, Barrington Jr. *Privacy: Studies in Social and Cultural History*. Armonk, N. Y.: M. E. Sharpe, 1984.

Nietzsche, Friedrich. *The Will to Power (1901)*. UK: Random House, 1968.

Nissenbaum, Helen. *Privacy in Context: Technology, Policy, and the Integrity of Social Life*. Stanford, CA: Stanford Law Books, 2010.

Plato. *Apology*.

-------*Laws*.

-------*Republic*.

Rabi, Lior. *Omes ha-individualiyut – ha-shorashim shel ideal ha-individualiyut ha-moderni* [The Burden of Individuality: The Origins of the Modern Ideal of Individuality] (in Hebrew). Haifa: Pardes, 2009.

Rakover, Nahum. *Hahaganah al Tzin'at Ha-prat* [Protection of Privacy in Jewish Law] (in Hebrew). Jerusalem: The Jewish Legal Heritage Society, 2006.

Rheingold, Howard. *The Virtual Community: Homesteading on the Electronic Frontier*. 2nd ed. Cambridge, Mass: MIT Press, 2000.

Robins, Kevin and Frank Webster. *Times of the Technoculture: From the Information Society to the Virtual Life*. London: Routledge, 1999.

Rokach, Lior and Oded Maimon. *Data Mining with Decision Trees: Theory and Applications*. 2nd edition. Singapore: World Scientific, 2014.

Rosen, Jeffrey. *The Unwanted Gaze: The Destruction of Privacy in America*. New York: Vintage Books, 2000.

Rousseau, Jean-Jacques. *The Social Contract*. Amsterdam, 1762.

Schneier, Bruce. *Data and Goliath: The Hidden Battles to Collect Your Data and Control Your World*. New York: Norton & Company, 2016.

Schoeman, Ferdinand. *Philosophical Dimensions of Privacy: An Anthology*. Cambridge: Cambridge University Press, 1984.

Simmel, Georg and Kurt H. Wolff. *The Sociology of Georg Simmel*. Glencoe: The Free Press, 1950.

Smith, Adam. *The Wealth of Nations*. London: W. Strahan and T. Cadell, 1776.

Solove, Daniel J. *Understanding Privacy*. London: Harvard University Press, 2009.

---------------------*Nothing to Hide*: The False Tradeoff between Privacy and Security. New Haven: Yale University Press, 2011.

Toffler, Alvin. *The Third Wave*. New York: Morrow, 1980.

Tonnies, Ferdinand and Charles P. Loomis. *Community & Society*. East Lansing: Michigan University Press, 1957.

Unguru, Sabetai. *Mavo le-toldot ha-matematika* [An Introduction to the History of Mathematics] (in Hebrew). Tel Aviv: Ministry of Defense Publications, 1989.

Westin, Alan. *Privacy and Freedom*. New York: Athenum, 1967.

Whiting, John and Irvin L. Child. *Child Training and Personality: A Cross-Cultural Study*. New Haven: Yale University Press, 1962.

Wiener, Norbert. *The Human Use of Human Beings: Cybernetics and
Society*. New York: Da Capo Press, 1954.

Zuboff, Shoshana. *The Age of Surveillance Capitalism*. London: Profile
Books, 2019.

Articles and Essays

Allcott, Hunt and Matthew Gentzkow. "Social Media and Fake News in
the 2016 Election". *Journal of Economic Perspectives* 31, no.
2 (Spring 2017): 211-236.

Alroy, Eran. "Ma ze, le-azazel, 'ha-Internet shel ha-dvarim'?" [What
the heck is 'the Internet of Things'?] (in Hebrew). *Anashim u
Makhshevim*, October 2014.
https://www.pc.co.il/editorial/168117 (Accessed 21/9/2020)

Amir, Asaf. "Kshe ha-pratiyut shelanu mekhulelet ba-reshet: al ha-
hashpaot ha-psychologiyot shel pish'ey ha-cyber" [When our
privacy is violated online: the psychological effects of
cybercrime] (in Hebrew). *Anashim u Makhshevim*, 10
February 2021. https://www.pc.co.il/featured/331923/
(Accessed 11/5/2021)

Armstrong, Mark, John Vickers and Jidong Zhou. "Consumer
Protection and the Incentive to Become Informed." *Journal of
the European Economic Association* 7 no. 2/3 (2009):
399-410.

Banton, Caroline. "Just in Time (JIT)." *Investopedia*.
https://www.investopedia.com/terms/j/jit.asp (Accessed
14/1/2021)

Barandiaran, Xabier E., Ezequiel Di Paolo and Marieke Rohde.
"Defining Agency: Individuality, Normativity, Asymmetry,

and Spatio-temporality in Action." In: *Adaptive Behavior* Vol. 17 (5) (October 2009): 367-386.

Barlow, John Perry. "The Economy of Ideas". *Wired Digital, Inc.*, 01/1994, 1-15. https://www.wired.com/1994/03/economy-ideas/ (Accessed 22/8/2020)

Ben-Israel, Isaac and Lior Tabansky. "Mabat beintkhumi al etgarey ha-bitakhon be-idan ha-meyda" [A multidisciplinary overview of security challenges in the information age] (in Hebrew). *Tzava ve-estrategia* Vol. 3 (December 2011): 19-32.

Benjamin, Garfield. "Privacy as a Cultural Phenomenon". *Journal of Media Critiques [JMC]* vol.3 No.10 (2017): 55-74.

Benoist, Emmanuel. "Collecting Data for the Profiling of Web Users". In: *Profiling the European Citizen: Cross-Disciplinary Perspectives*, eds. Mireille Hildebrandt and Serge Gutwirth, 169-184. Dordrecht: Springer, 2008.

Campbell, James, Avi Goldfarb and Catherine Tucker. "Privacy Regulation and Market Structure". *Journal of Economics and Management Strategy* 24 no.1 (2015): 47-73.

Cate, Fred H., Peter Cullen and Viktor Mayer-Schönberger. "Data Protection Principles for the 21st Century". Books and book chapters by Maurer faculty, Indiana University, 2013.

Chappell, David. "A short introduction to cloud platforms: An enterprise-oriented view." http://www.davidchappell.com/CloudPlatforms--Chappell.pdf (Accessed 29/7/2024)

Confessore, Nicholas. "Cambridge Analytica and Facebook: The Scandal and the Fallout So Far". *The New York Times*, 4 April 2018. https://www.nytimes.com/2018/04/04/us/politics/cambridge-analytica-scandal-fallout.html (Accessed 5/4/2024)

Dollarhide, Maya E. "Mass Customization." *Investopedia.*
https://www.investopedia.com/terms/m/masscustomization.asp
(Accessed 14/1/2021)

EU GDPR – Article 22. https://gdpr-info.eu/art-22-gdpr/ (Accessed
1/7/2021)

EU GDPR – Recital 71. https://gdpr-info.eu/recitals/no-71/ (Accessed
1/7/2021)

Evans Grubbs, Judith. "Making the Private Public: Illegitimacy and
Incest in Roman Law." In: *Public and Private in Ancient
Mediterranean Law and Religion*, eds. Jörg Rüpke and
Clifford Ando, 115-141. Berlin: De Gruyter, 2015.

Fisher, Eran. "Lekhudim ba-reshet: siakh ha-technologiya ha-rishtit ve
ha-capitalism ha-khadash" [Caught in the net: the new
capitalism and network technology discourse] (in Hebrew).
Teoriya ve Bikoret 37 (Fall 2010): 150-183.

Floridi, Luciano. "The Informational Nature of Personal Identity".
Minds and Machines Vol.21 (4) (November 2011): 549-566.

--------------------"Philosophical Conceptions of Information." In:
*Formal Theories of Information: From Shannon to Semantic
Information Theory and General Concepts of Information*, ed.
Giovanni Sommaruga, 13-53. Springer, 2009.

Fox, Chris. "Google and Facebook accused of breaking GDPR laws."
BBC Online, 25 May 2018.
https://www.bbc.com/news/technology-44252327 (Accessed
12/5/2024)

Friedman, Arik and Schuster Assaf. "Data Mining with Differential
Privacy". Delivered at: Privacy in Challenging Times: The 8th
Technion Summer School on Cyber and Computer Security,

September 2020.

Gavison, Ruth. "Privacy and the Limits of Law". *The Yale Law Journal,* Vol. 89, No. 3 (January 1980): 421-471.

Gerstein, Robert S. "Intimacy and Privacy." *Ethics*, Vol.89 (1) (October 1978): 76-81.

Glickman, Aviad. "Parashat mirsham ha-okhlosin: Kakh dalaf ha-meyda la-reshet" [The Civil Registry Breach Affair: This is how the information was leaked out]. *Ynet*, 24 October 2011. https://www.ynet.co.il/articles/0,7340,L-4138213,00.html (Accessed 15/07/2021)

Godkin, E. L. "Libel and Its Legal Remedy." *Journal of Social Science* 12 (December 1880): 69-80.

Gotlieb, Calvin C. "Privacy: A Concept Whose Time Has Come and Gone." In: *Computer, Surveillance, and Privacy*, eds. David Lyon and Elia Zuriek, 71-156. Minneapolis: University of Minnesota Press, 1996.

Gradwohl, Ronen and Rann Smorodinsky. "Subjective Perception Games and Privacy". 8 September 2014. https://www.researchgate.net/publication/265385815_Subjective_Perception_Games_and_Privacy (Accessed 1/07/2021)

Hansen, Marit and Andreas Pfitzmann. "A terminology for talking about privacy by data minimization: Anonymity, Unlinkability, Undetectability, Unobservability, Pseudonymity, and Identity Management". https://www.researchgate.net (Accessed 24/1/2021)

Hildebrandt, Mireille. "Privacy as Protection of the Incomputable Self: From Agnostic to Agonistic Machine Learning." *Theoretical Inquiries in Law,* vol.20 (1) (2019): 83-121.

--------------------------"Defining Profiling: A New Type of

Knowledge?". In: *Profiling the European Citizen: Cross-Disciplinary Perspectives*, eds. Mireille Hildebrandt and Serge Gutwirth, 17-45. Dordrecht: Springer, 2008.

Hodge, Neil. "Ireland GDPR caseload nearly doubled in 2019". *Compliance Week*, 20 February 2020. https://www.complianceweek.com/data-privacy/ireland-gdpr-caseload-nearly-doubled-in-2019/28475.article (Accessed 12/5/2024)

IBM website. "What is deep learning?" https://www.ibm.com/topics/deep-learning (Accessed 12/3/2024)

Information Commissioner's Office (ICO). "Data Protection and the EU". https://ico.org.uk/for-organisations/data-protection-and-the-eu/ (Accessed 20/6/2024)

Internet Encyclopedia of Philosophy. "Autonomy". https://iep.utm.edu/autonomy/ (Accessed 28/3/2024)

Jank, Werner. "Didaktik, Bildung, Content: On the Writings of Frede V. Nielsen". *Philosophy of Music Education Review* vol.22 (2) (Fall 2014): 113-131.

Jessop, Bob. "Fordism and Post-Fordism: A Critical Reformulation." In: *Pathways to Industrialization and Regional Development*, eds. Michael Storper and Aleen J. Scott, 1-41. London: Taylor & Francis, 1992.

Kagal, Lalana and Hal Abelson. "Access Control is an Inadequate Framework for Privacy Protection." MIT Computer Science and Artificial Intelligence Lab, 2010.

Konvitz, Milton R. "Privacy and the Law: A Philosophical Prelude". *Law and Contemporary Problems* 31 (2) (Spring 1966): 272-280.

Laney, Douglas. "Application Delivery Strategies". *Meta Group*,
February 2001.
https://www.academia.edu/37216189/Application_Delivery_S
trategies (Accessed 30/1/2021)

Levin, Ilya. "Cultural trends in a digital society". *Proceedings of TMCE
2014 (19-23 May 2014), Budapest*, eds. I. Horváth and Z.
Rusák: 13-21.

----------------"Cyber-physical Systems as a Cultural Phenomenon."
International Journal of Design Sciences and Technology
Vol.22 (2016): 67-80.

Li, Ninghui, Tiancheng Li and Suresh Venkatasubramanian.
"t-Closeness: Privacy Beyond k-Anonymity and l-Diversity."
2007 IEEE 23rd International Conference on Data
Engineering, 5-20 April 2007.

Lukács, Adrienn. "What is Privacy? The History and Definition of
Privacy". 2016.
https://www.academia.edu/31980162/What_is_Privacy_The_
History_and_Definition_of_Privacy (Accessed 17/6/2024)

Lyon, David. "From 'Post-Industrialism' to 'Information Society': A
New Social Transformation?". *Sociology* vol.20 (4)
(November 1986): 577-588.

MacAskill, Ewan and Julian Borger. "New NSA leaks show how the
US is bugging its European allies." *The Guardian*, 30 June
2013. https://www.theguardian.com/world/2013/jun/30/nsa-
leaks-us-bugging-european-allies (Accessed 5/4/2024)

Machanavajjhala, Ashwin, Johannes Gehrke and Daniel Kifer.
"l-Diversity: Privacy Beyond k-Anonymity." 22nd
International Conference on Data Engineering (ICDE'06), 3-7
April 2006.

Marx, Gary T. "Varieties of Personal Information as Influences on
Attitudes Toward Surveillance." In: *The New Politics of
Surveillance and Visibility*, eds. Kevin Haggerty and Richard
Ericson, 79-110. Toronto: University of Toronto Press, 2006.

Marx, Karl and Friedrich Engels. "Manifesto of the Communist Party"
(1848). *Marxists Internet Archive*
https://www.marxists.org/archive/marx/works/download/pdf/
Manifesto.pdf (Accessed 16/6/2024)

McDonald, Aleecia and Lorrie Faith Cranor. "The Cost of Reading
Privacy Policies". *A Journal of Law and Policy for the
Information Society*, 4, no. 3 (2008): 543-568.

Mollick, Ethan. "Establishing Moore's Law". *IEEE Annals of the
History of Computing* vol. 28, no. 3 (July 2006): 62-75.

Moore, Adam D. "Privacy: Its Meaning and Value". *American
Philosophical Quarterly*, vol. 40, no. 3 (July 2003): 215-227.

Negley, Glenn. "Philosophical views on the value of privacy." *Law and
Contemporary Problems* 31 (Spring 1966): 319-325.

New York Times staff. "Zuckerberg and Dorsey Face Harsh
Questioning From Lawmakers." *The New York Times*, 17
November 2020.
https://www.nytimes.com/live/2020/11/17/technology/twitter-
facebook-hearings (Accessed May 27, 2024)

Nielsen, Frede V. "Music (and Arts) Education from the Point of View
of Didaktik and Bildung". In: *International Handbook of
Research in Arts Education*, ed. Liora Bresler, 265-285.
Dordrecht: Springer, 2007.

Nissenbaum, Helen. "The Meaning of Anonymity in an Information
Age". *The Information Society* Vol.15 (2) (May 1999):
141-144.

Nissim, Kobbi. "Differential Privacy: Why, How and Where to?".
    Delivered at: Privacy in Challenging Times: The 8th Technion
    Summer School on Cyber and Computer Security (September
    2020).

Parker, Richard B. "A Definition of Privacy*". *Rutgers Law Review*
    Vol. 27, Issue 2 (1974): 275-297.

Periñán, Bernardo. "The Origin of Privacy as a Legal Value: A
    Reflection on Roman and English Law." *American Journal of
    Legal History*, vol. 52 (2) (April 2012): 183-201.

Pilot, Adrian. "Ma Pfizer omedet lada'at aleynu?" [What is Pfizer about
    to know about us?] (in Hebrew). *Calcalist*, 10 January 2021.
    https://www.calcalist.co.il/local/articles/0,7340,L-
    3886752,00.html (Accessed 15/07/2021)

Popper, Karl. "Three Worlds – The Tanner Lecture on Human Values".
    Delivered at the University of Michigan, April 7, 1978.

Posner, Richard A. "A theory of primitive society, with special
    reference to law." *The Journal of Law & Economics* vol.23 (1)
    (April 1980): 487-501.

Quinlan, J. R. "Induction of Decision Trees". *Mach. Learn.* 1, 1 (Mar.
    1986): 81-106.

Rodogno, Raffaele. "Personal Identity Online". *Philosophy &
    Technology* Vol. 25 (3) (2012): 309-328.

Satariano, Adam. "Google Fined $1.7 Billion by E.U. for Unfair
    Advertising Rules". *The New York Times*, 20 March 2019.
    https://www.nytimes.com/2019/03/20/business/google-fine-
    advertising.html (Accessed 27/5/2024)

Schafer, Arthur. "Privacy: A Philosophical Overview". In: *Aspects of
    Privacy Law: essays in honor of John M. Sharp*, ed. Dale
    Gibson, 1-20. Toronto: Butterworths 1980.

Schechtman, Marya. "Stories, Lives, and Basic Survival: A Refinement
and Defense of the Narrative View." In: *Narrative and
Understanding Persons*, ed. Daniel Hutto, 155-178.
Cambridge: Cambridge University Press, 2007.

Shakhaf, Tal. "2020 hayta shana shkhora la-pratiyut be-Israel" [2020
was a dark year for privacy in Israel] (in Hebrew). *Ynet*, 29
January 2021.
https://www.ynet.co.il/digital/technology/article/ryAzJe11xu
(Accessed 11/5/2021)

Shalev, Eran. "Ha-masekhot noflot: aliyata shel ha-democratia ha-
americayit ve-ovdan ha-pratiyut ba-siyakh ha-politi" [The
Rise of Democracy and the Decline of Republican Privacy in
the United States] (in Hebrew). *Zmanim: A Historical
Quarterly*, 104 (Fall 2008): 34-43.

Shannon, Claude E. "A Mathematical Theory of Communication".
Reprinted with corrections from *The Bell System Technical
Journal* vol. 27 (October 1948).

Simmel, Georg. "How is Society Possible?". *American Journal of
Sociology* Vol. 16(3) (November 1910): 372-391.

Solove, Daniel J. "Conceptualizing Privacy." *California Law Review*
Vol. 90 (4) (July 2002): 1088-1154.

Stanford Encyclopedia of Philosophy. "Semantic Conceptions of
Information." First published Wed Oct 5, 2005; substantive
revision Wed Jan 7, 2015.

Sweeney, Latanya. "*k*-Anonymity: A Model for Protecting Privacy."
*International Journal on Uncertainty, Fuzziness and
Knowledge-based Systems*, 10 (5) (October 2002): 557-570.

Travers, Jeffrey and Stanley Milgram. "An Experimental Study of the Small World Problem". *Sociometry* vol. 32, 4 (Dec. 1969): 425-443.

Turing, A. M. "Computing Machinery and Intelligence". *Mind* 49 (1950): 433-460.

Vieira, Armando. "Predicting online user behaviour using deep learning algorithms." 10 February 2016. https://www.researchgate.net/publication/284219029_Predicting_online_user_behaviour_using_deep_learning_algorithms

Wahrman, Dror and Yaniv Farkash. "Im ein ani li: Ha-prat ha-naor be-idan ha-mahapekhot" [The Enlightenment and Selfhood: A New Perspective on the Age of Revolutions] (in Hebrew). *Zmanim: A Historical Quarterly*, 104 (Fall 2008): 62-71.

Warren, Samuel D. and Louis D. Brandeis. "The Right to Privacy" (1890), in: *Philosophical Dimensions of Privacy: An Anthology*, ed. Ferdinand D. Schoeman, 75-103. Cambridge: Cambridge University Press, 1984.

Westin, Alan. "The origins of modern claims to privacy." In: *Philosophical Dimensions of Privacy: An Anthology*, ed. Ferdinand D. Schoeman, 56-59. Cambridge: Cambridge University Press, 1984.

Wheeler, John Archibald. "Information, Physics, Quantum: The Search for Links". In: *Complexity, Entropy, and the Physics of Information*, ed. Wojciech H. Zurek, 309-336. CA: Westview Press, 1990.

Yakobson, Alexander. "Ma mevin nagar be-politica? Ha-democratia ve-hatzdakata be-mashal Protagoras" [What Does a Carpenter Know About Politics?: Democracy and its Justification in

Plato's 'Protagoras'] (in Hebrew). *Zmanim: A Historical Quarterly*, 64 (Fall 1988): 23-33.

Appendices

Appendix A

Game Theory Models

The balance between individuals' interest in protecting their private information and the interests of other entities (other individuals, confidants, Internet companies, corporations and government agencies) has been disrupted in the age of ICTs.[554] To find new points of balance, I suggest the use of game theory models, which I will describe now.

Why Use Game Theory Models

In the age of ICTs, our personal information has become a commodity in the hands of players in the commodity market.[555] Those players use individuals' personal information to maximize their profits. The said individuals, being nodes in the network, seemingly give out their private personal information for free; they provide it every time they make an online purchase, receive medical treatment, pay their taxes, search information on Google or just browse the web for fun. This state of events justifies the use of game theory models (exemplars) to understand the rationale behind network players' decisions to violate personal

---

[554] For a discussion of the cracks in the current paradigm of personal privacy, see Chapter 11.

[555] Lisa Rajbhandari and Einar Arthur Snekkenes, "Using Game Theory to Analyze Risk to Privacy: An Initial Insight", presented at IFIP PrimeLife International Summer School on Privacy and Identity Management for Life, 2010, 42.

privacy. That understanding will allow us to come up with alternatives that may affect the players' considerations and lead to reduced violation of personal privacy. **For each model I present here, I will describe possible alternatives that may be able to prevent or reduce personal privacy violations.**

Terms and Theorems

- The players are selfish and rational: They act out of rational considerations of gain and loss and solely to maximize their payoff.
- Perfect information game: At every stage of the game, every player has full knowledge of all the stages of the game, of all the players' previous moves, and of the strategies each player may use during the game.
- Non-cooperative game: A game in which players cannot form alliances or agree on solutions (equilibria). Each player is acting selfishly and rationally to maximize its payoff.
- Strategy: Players' strategy is what guides their actions and choices throughout the game, telling them what to do in every possible situation.[556]
- Nash equilibrium: A situation where no player can gain by changing its strategy (the strategy vector $S^* = (s_1^* \ldots s_i^* \ldots s_n^*)$ is a Nash equilibrium if sticking with s_i^* is the best option for player i).[557]

---

[556] Ibid., 43.

[557] Michael Maschler, Eilon Solan and Shmuel Zamir, *Game Theory* (New York: Cambridge University Press, 2013), 97.

- Dominant strategy: A player's strategy that is equal or superior to every other strategy available to that player.[558]

- Pure and mixed strategies: A pure strategy is any strategy S_i used by player i. A mixed strategy is the assignment of a probability between *0* and *1* to each of the pure strategies available to player i – let us call them *j*, *m*, and *n* – so that $p_j + p_m + p_n = 1$.[559]

Setting and Moves

The move options – to share or to withhold personal information – are based on each of the players' possession of said personal information. The information possession options are:

- Information possessed solely by the individual themself (deep personal privacy);
- Information possessed by the individual and shared by them with a confidant (general personal privacy);
- Information possessed by others who are not confidants (personal information that is known publicly);
- Information possessed solely by the confidant and not by anyone else (the little hump);

---

[558] Ibid., 105.

[559] Robert Aumann, Yair Tauman and Shmuel Zamir, *Torat ha-miskhakim – Yekhidot 1, 2, 3* [Game *Theory – Units 1, 2 & 3*] (in Hebrew) (Tel Aviv: Everyman's University, 1981), 19.

- Information possessed by the public but not by the individual in question (the big hump).

This can be graphically represented as follows:

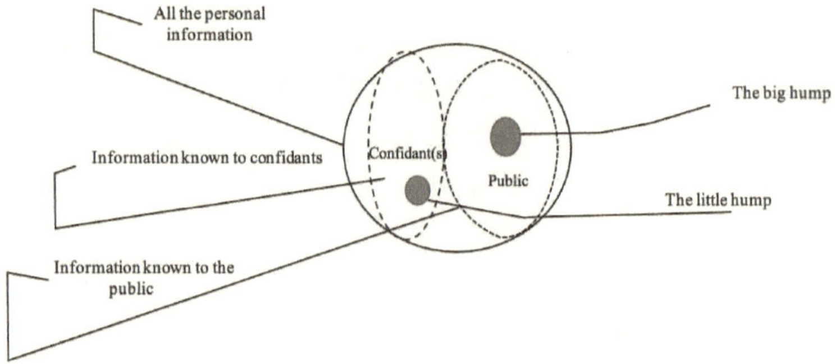

The outermost oval represents all the personal information about an individual, the dashed oval represents the personal information known to confidants, and the dotted oval represents the personal information that is known to the general public.

Deep personal privacy information (all the information known only to the person in question) is all the area outside the dashed and dotted ovals.

General personal privacy information is the entire area of the outermost oval minus the area of the dotted oval.

Exposure of personal privacy information creates the "humps" – information about the individual possessed by confidants and/or the general public but not by the individual themself. On the graph, the humps are represented by the two smaller ovals inside the dotted and dashed ovals.

Basic Conditions

a. The game has three players:

- An individual (Player I) who possesses their deep privacy information;
- A confidant (Player II) who possesses Player I's general privacy information and "little hump" information;
- Internet companies (Player III), which possess Player I's information that is publicly known as well as "big hump" information.

b. The set prices are as follows:

- The free market price of a unit of information is x;
- The fine for violating Player I's privacy is y.[560]

c. The game may have one or more rounds. Each round is independent of the previous ones, and therefore, in each round, a player may only choose one of the options available.

d. For the sake of convenience, let the aforementioned information possession options be numbered as follows:

- Information possessed solely by the individual – 1
- Information possessed by the individual and shared by them with a confidant – 2
- Information possessed by others who are not confidants – 5
- Information possessed solely by the confidant – 3

---

[560] In this game, only Player I's privacy can be violated.

- Information possessed by the public but not by the individual in question – 4

e. The basic payoff matrix of the game:[561]

Information source	Information destination	Player I			Player II			Player III		
		Value of privacy loss	Gain for privacy loss	Total	Fine for source node privacy loss	Gain for source node privacy loss	Total	Value of source privacy loss	Gain for source privacy loss	Total
1	2	$-I_{12} \cdot x$	0	$-I_{12} \cdot x$	$-I_{12} \cdot y$	$I_{12} \cdot x$	$I_{12} \cdot x - I_{12} \cdot y$			
1	5	$-I_{15} \cdot x$	0	$-I_{15} \cdot x$					$I_{15} \cdot x$	$I_{15} \cdot x$
2	5				$-I_{15} \cdot y$	$I_{15} \cdot x$	$I_{25} \cdot x - I_{25} \cdot y$		$I_{15} \cdot x$	$I_{15} \cdot x$
3	1	$I_{31} \cdot x$	0	$I_{31} \cdot x$		$I_{31} \cdot x$	$I_{31} \cdot x$			
3	4				$-I_{34} \cdot y$	$I_{34} \cdot x$	$I_{34} \cdot x - I_{34} \cdot y$		$I_{34} \cdot x$	$I_{34} \cdot x$
4	1	$I_{34} \cdot x$	0	$I_{34} \cdot x$					$I_{34} \cdot x$	$I_{34} \cdot x$

This matrix reflects the actual current state of events concerning online privacy and means the following:

- The individual (Player I) gives up their privacy for free – unless we consider the services they get from their confidants, and especially from Internet companies (in the form of "free" apps), to be considered.

- The confidant (Player II) may gain from using Player I's private information but may be fined for selling elements of Player I's private information, which would be a violation of its ethical and legal obligations to protect the confidentiality of Player I's personal information.

- Player III (Internet companies) may freely trade in individuals' private information. This

---

[561] The payoff matrices of the specific game models that follow are based on this general matrix.

assumption is in line with the analysis of surveillance capitalism presented in Chapter 9.

- One way to balance out the interests of the different players in this game would be to fine Internet companies for privacy violations, as stipulated in the GDPR.

- All the values in the payoff matrices discussed in this appendix are probable examples given based on subjective assessment. To learn the actual value of personal information for the different players included in this matrix, future researchers may survey players from the different categories to obtain their assessment of the payoffs in the models provided here. My underlying assumption, without loss of generality, was that all the basic components of privacy have the same value for Player I and the same (though different from Player I's) value for Player II.

Based on this general payoff matrix, we can define three equilibrium games:

1. Individual vs confidant
2. Individual vs. Internet company
3. Confidant vs Internet company

Game Model 1: Individual vs Confidant

The personal information protection game scenario between an individual and their confidant is as follows:

The individual and the confidant both possess the individual's private personal information. The individual shares this information with the confidant to enjoy its services (e.g., medical care, legal counseling, financial counseling, or customized recommendations for products). The confidant makes a commitment to protect the information and not to share it with any third parties – a commitment that is often enforced by law (as in the case of doctor-patient confidentiality or attorney-client privilege) or at least required by social norms (e.g., sharing of intimate information between sexual partners). Nevertheless, the confidant is tempted to trade the individual's personal information with third parties because, as I have shown, personal information is a valuable resource that may bring the confidant considerable profit. In view of this relationship, let us analyze the "game" between the two, where each party chooses the strategy that best serves its interests and looks for equilibria.

The Game Setting

For the basic conditions of the game, see the section "Basic Conditions."

Players: Player I is an individual who possesses private personal information;[562] Player II is a confidant with whom Player I sees a candidate for sharing their private personal information.[563]

---

[562] For the sake of this example, let Player I be a patient.
[563] Let Player II be a doctor.

Available **strategies:** Player I (the individual) has two possible strategies: to share their information in full or to share it in part.[564] Player II (the confidant) also has two possible strategies: to maintain the confidentiality of Player I's information or to trade it.[565]

The Game's Payoff Matrix

Let the general payoff matrix be:

	Information value for Player I		Information value for Player II	
	Player I shares full information.	Player I shares partial information	Player I shares full information	Player I shares partial information
Player II maintains information confidentiality	(a) 1	(b) 0.5	(c) 1	(d) 0.5
Player II sells the information	(e) -1	(f) -0.1	(g) 0.9	(h) 0.2

---

[564] The motive for sharing full information is clear – the desire to receive better service and help from the confidant (e.g., medical care). Choosing to share only part of the information may be based on a belief that it will better serve your interests: e.g., if you do not disclose all your financial information, you may be able to get better loan terms.

[565] Choosing to maintain confidentiality usually stems from legal or normative obligation, while choosing to sell the information is usually motivated by the desire to make a profit of it.

Payoff matrix justification:

For Player I:

If Player II maintains the confidentiality of Player I's information, Player I will benefit the most from sharing full information with Player II: in our example, it will allow the doctor Player II to provide optimal medical care to the patient Player I. Therefore, the information value is *1*. If Player I shares only part of their information, they will receive care that is less than optimal, and therefore, the information value is only *0.5*.

If Player II violates their confidentiality commitment, Player I will suffer harm (e.g., their insurance company will raise their life insurance premium as a result) at a value of *-1* if they shared their full information or *-0.1* if they only shared part of their information.

For Player II:

If Player II received full information from Player I, maintaining confidentiality allows them to fully realize their abilities (in our example, provide the best medical care, which would earn them a good reputation); therefore, the information value is *1* for them as well. If they received partial information, the value of that information is only *0.5*.

Violating Player I's confidentiality in case of possessing full information gives Player II a value of *0.9* (the value of selling the information). In case of partial information, the value of selling is only *0.2*.

Importantly, when Player II is choosing a strategy, they cannot know whether Player I has given them full or partial information. Only Player I can know that.

Based on this, we can calculate a more specific payoff matrix for each of the players. Let us mark Player I's payoffs as X_{ij}, with $i = 1$ meaning provision of full information by Player I, $i = 2$ meaning provision of partial information, $j = 1$ meaning maintaining of confidentiality by Player II, and $j = 2$ meaning breach of confidentiality. Similarly, let us mark Player II's payoffs as Y_{ij}. **I am assuming a confidant will always sell the information it has received to maximize its profit.**[566]

The game payoffs will be as follows:

$X_{11} = (a) = 1$

$X_{12} = (a) + (e) = 1 + (-1) = 0$

$X_{21} = (b) = 0.5$

$X_{22} = (b) + (f) = 0.5 + (-0.1) = 0.4$

$Y_{11} = (c) = 1$

$Y_{12} = (c) + (g) = 1 + 0.9 = 1.9$

$Y_{21} = (d) = 0.5$

$Y_{22} = (d) + (h) = 0.5 + 0.2 = 0.7$

For convenience, let us display this payoff matrix in tabular form:

Player I's payoff matrix:

	Player I shares full information	Player I shares partial information
Player II maintains confidentiality	1	0.5
Player II breaches confidentiality	0	0.4

Player II's payoff matrix:

	Player I shares full information	Player I shares partial information
Player II maintains confidentiality	1	0.5
Player II breaches confidentiality	1.9	0.7

The game has a pure strategy Nash equilibrium. Player II has the dominant strategy: whatever the case, breaching Player I's confidentiality pays off. If Player I chooses the strategy of sharing full information, and Player II maintains its confidentiality, both will have a payoff of *1*. However, breaching confidentiality will increase Player II's payoff to *1.9*. Therefore, the best strategy for Player I will be to initially share only partial information to secure a payoff of at least *0.4*. But even then, it still pays off for Player II to breach confidentiality, as their payoff will be *0.7* instead of *0.5*. Since neither player can fare better than that, this is the Nash equilibrium of this game.[567]

This is an example of a one-time game in which distrust between an individual and their confidant leads to both of them not getting the optimal outcome they might have had if they had maintained general privacy. The cost of privacy violation is *1 – 0.4 = 0.6* for Player I and *1 – 0.7 = 0.3* for Player II.

We can assume there will also be a social cost for violating privacy in this game: *0.6 + 0.3 = 0.9*. This supports the argument that maintaining personal privacy serves society even more than it serves the individual.

---

[567] See "Appendix A – Summary and Conclusions" for suggested ways to improve this state of events.

Game Model 2: Individual vs Internet Company

In this game scenario, the individual and the Internet company both possess the individual's private personal information. The individual shares this information with the Internet company (a navigation app like Waze, a social network like Twitter, or a search engine like Google) to enjoy the services it provides. The Internet company makes a commitment to protect its users' information privacy and usually asks its users to affirm they have read and agree to the company's privacy policy. The Internet company has an obligation to comply with its declared information security policy; however, it is sorely tempted to trade the individual's personal information with third parties using loopholes in the said policy because, as explained earlier, personal information is a valuable resource that can bring it considerable profit. In view of this relationship, let us analyze the "game" between the two, where each party chooses the strategy that best serves its interests and looks for equilibria.

The Game Setting

The basic conditions are the same as in game model 1.

Players: Player I is an individual who possesses private personal information.[568] Player II is an Internet company (for this example, let it be the company that owns Waze) whose Internet-based service the individual seeks to use.

---

[568] For the sake of this example, let Player I be a driver who wants to get from A to B.

Available **strategies:** Player I (the individual) has two possible strategies: to share their full information or to share only the minimal information[569] that is necessary for using the service.[570] Player II (the Internet company) also has two possible strategies: to keep the information to itself or to trade it.[571]

---

[569] The minimal information that the Internet company needs in order to provide its declared service. In the case of Waze, it is the driver's phone number, the required destination, and Internet connection on the driver's phone.

[570] Reasons for choosing to share full information may include Player I's indifference to the ways their personal information may be used (as discussed in Chapters 9-10), or unawareness of the possibilities of improper use of their personal information. Choosing to share only the required minimum may be based on a belief that it will better serve the player's interests: e.g., if you do not disclose all your financial information, you may be able to get better loan terms.

[571] Choosing to maintain confidentiality usually stems from legal or normative obligation, while choosing to sell the information is usually motivated by the desire to make a profit of it.

The Game's Payoff Matrix

Let the general payoff matrix be:

	Information value for Player I		Information value for Player II	
	Player I shares full information	Player I shares minimal information	Player I shares full information	Player I shares minimal information
Player II maintains information confidentiality	(a)	(b)	(c)	(d)
	1.2	1	1.2	0.7
Player II sells the information	(e)	(f)	(g)	(h)
	-1	-0.5	0.9	0.2

Payoff matrix justification:

For Player I:

If Player II maintains the confidentiality of Player I's information, Player I will benefit the most from sharing full information with Player II: in our example, Player II (Waze) can give the driver Player I customized recommendations for stops along their route based on Player I's shopping preferences. Therefore, the information value is *1.2*. If Player I shares only the minimal information required, Player II will be able to provide exactly its declared service and nothing more. Therefore, the information value for Player I, in this case, is *1*.

If Player II violates its confidentiality commitment, Player I will suffer harm (e.g., the confidentiality breach will reveal who the driver has spoken to during the trip) at a value of *-1* if they shared their full information or *-0.5* if they shared only the minimal information.

For Player II:

If Player II received full information from Player I, maintaining confidentiality allows it to fully realize its capabilities (in our example, Waze can offer extra services, such as customized recommendations for stops based on the driver's preferences), and therefore, the information value is *1.2* for it as well. If Player II received only the minimal information it needs to provide the service, the value of that information for Player II is *0.7*.

Violating Player I's confidentiality in case of possessing full information gives Player II a value of *0.9* (the value of selling the information). In case of partial information, the value of selling is only *0.2*.

Importantly, when Player II is choosing a strategy, it cannot know whether Player I has given it full or partial information. Only Player I can know that.

Based on this, we can calculate a more specific payoff matrix for each of the players. Let us mark Player I's payoffs as X_{ij}, with $i = 1$ meaning provision of full information by Player I, $i = 2$ meaning provision of partial information, $j = 1$ meaning maintaining of confidentiality by Player II, and $j = 2$ meaning violation of confidentiality. Similarly, let us mark Player II's payoffs as Y_{ij}. **I am assuming Player II will always not only use the personal information it has received from Player I but also sell it to maximize its profit.**[572]

The game payoffs will be as follows:

---

[572] As described in Chapter 9, in the section on surveillance capitalism.

$X_{11} = (a) = 1.2$

$X_{12} = (a) + (e) = 1.2 + (-1) = 0.2$

$X_{21} = (b) = 1$

$X_{22} = (b) + (f) = 1 + (-0.5) = 0.5$

$Y_{11} = (c) = 1.2$

$Y_{12} = (c) + (g) = 1.2 + 0.9 = 2.1$

$Y_{21} = (d) = 0.7$

$Y_{22} = (d) + (h) = 0.7 + 0.2 = 0.9$

For convenience, let us display this payoff matrix in tabular form:

Player I's payoff matrix:

	Player I shares full information	Player I shares partial information
Player II maintains confidentiality	1.2	1
Player II breaches confidentiality	0.2	(0.5)

Player II's payoff matrix:

	Player I shares full information	Player I shares partial information
Player II maintains confidentiality	1.2	0.7
Player II breaches confidentiality	2.1	(0.9)

The game has a pure strategy Nash equilibrium. Player II has the dominant strategy: whatever the case, breaching Player I's confidentiality pays off. If Player I chooses the strategy of sharing full information and Player II maintains its confidentiality, both will have a payoff of *1.2*. However, breaching confidentiality will increase Player II's payoff to *2.1*. Therefore, the best strategy for Player I will be to initially share only partial information to secure a payoff of at least *0.5*. But even then, it still pays off for Player II to breach confidentiality, as their payoff will be *0.9* instead of *0.7*. Since neither player can fare better than that, this is the Nash equilibrium of this game.[573]

This is an example of a one-time game of prisoner's dilemma, in which distrust between the individual and the Internet company leads to both of them not getting the optimal outcome they might have had if they had maintained general privacy. The cost of privacy violation is *1.2 – 0.5 = 0.7* for Player I, and *1.2 – 0.9 = 0.3* for Player II.

We can assume there will also be a social cost for violating privacy in this game: *0.7 + 0.3 = 1*. This supports the argument that maintaining personal privacy serves society even more than it serves the individual.

Game Model 3: Confidant vs Internet Company

The personal information protection game scenario between a confidant entity[574] and an Internet company[575] is as follows:

---

[573] See "Appendix A – Summary and Conclusions" for suggested ways to improve this state of events.
[574] An entity that serves as a confidant to an individual.
[575] E.g., Google.

In this game, each of the players possesses some private personal information of a given individual. Both received that information in exchange for their services. It does not matter whether the information is full or partial. Both players are committed to maintaining the individual's privacy – whether by law, by social norms, and/or by their own privacy policy. Both are sorely tempted to trade the personal information they possess, as it can bring them considerable profit; however, they may be fined if they breach their privacy obligations.

In view of this relationship, let us analyze the "game" between the two, where each party chooses the strategy that best serves its interests and looks for equilibrium points.

The Game Setting

The basic conditions are the same as in game models 1 and 2.

Players: Player I is a confidant, i.e., an entity in which individuals see a candidate for sharing their private personal information with;[576] Player II is an Internet company, e.g., Google.[577]

Available **strategies:** Player I (the confidant) has two possible strategies: to protect the individual's privacy or to trade their private information. Player II (the Internet company) also has two possible strategies: to protect the confidentiality of the information it possesses or to trade it.[578]

---

[576] For the sake of this example, let Player I be a health insurance company whose representatives search medical information about individuals online.

[577] The search engine where Player I is looking for medical information.

[578] Choosing to maintain confidentiality usually stems from legal or normative obligation, while choosing to sell the information is usually

The Game's Payoff Matrix

Let the general payoff matrix be:

	Information value for Player I		Information value for Player II	
	Player I maintains confident iality	Player I breaches confident iality	Player I maintains confident iality	Player I breaches confident iality
Player II maintains confident iality	(a) 1	(b) -0.3	(c) 0.3	(d) 0.4
Player II breaches confident iality	(e) 0.2	(f) 0.9	(g) -0.2	(h) 1.5

Payoff matrix justification:

For Player I:

If both Player I and Player II maintain the confidentiality of the personal information in their possession, Player I has no additional gain beyond the gain it already has from being a confidant; therefore, the value of the information for it is *1*. If Player I maintains confidentiality, but Player II does not, Player I has an additional gain of *0.2* from obtaining new information it did not initially have. If Player I does not

---

motivated by the desire to make a profit of it.

maintain confidentiality, but Player II does, Player I profits from selling the information in its possession but may be fined for privacy violations,[579] which makes its total payoff negative – *-0.3*. If both players breach confidentiality, Player I gains more from obtaining new information it did not initially have, which outweighs any potential fines and leaves it with a total payoff of *0.9*.

For Player II:

If both Player I and Player II maintain confidentiality, Player II gains only from the information it has in its possession; however, that gain should be cut off against the costs of the service it provides to the user "for free," which makes its total payoff *0.3*. If Player II maintains confidentiality, but Player I does not, Player II gains from obtaining new information from Player I and is not subject to any fines itself, meaning an additional gain of *0.4*. If Player II does not maintain confidentiality, but Player I does, Player II profits from selling the information in its possession. It may be fined at a value of *0.2* for privacy violations,[580] but the fine is outweighed by the profit it gains from selling the information. If both players breach confidentiality, Player II has additional gain from obtaining new information it did not previously have and is at lesser risk of being fined: when an individual's confidentiality is breached by more than one entity, it is harder to find who is guilty; both players can play a "blame game" and eventually get away with virtually no punishment. Therefore, in this case, Player II's payoff is *1.5*. Thus, **if both players breach individuals' confidentiality and simultaneously deny it, Player II gains the most** because it possesses more personal information that it can trade, as we saw in the discussion of surveillance capitalism.

---

[579] If it is bound by privacy protection laws. It is also at risk of being reported by Player II.

[580] E.g., if it violates the GDPR.

Based on this, we can calculate a more specific payoff matrix for each of the players. Let us mark Player I's payoffs as X_{ij}, with $i = 1$ meaning maintaining confidentiality and i = 2 meaning violation of confidentiality by Player I, and with j = 1 meaning maintaining confidentiality and $j = 2$ meaning violation of confidentiality by Player II. Similarly, let us mark Player II's payoffs as Y_{ij}.

The game payoffs will be as follows:

$X_{11} = (a) = 1$

$X_{12} = (a) + (e) = 1 + 0.2 = 1.2$

$X_{21} = (b) = -0.3$

$X_{22} = (b) + (f) = -0.3 + 0.9 = 0.6$

$Y_{11} = (c) = 0.3$

$Y_{12} = (c) + (g) = 0.3 - 0.2 = 0.1$

$Y_{21} = (d) = 0.4$

$Y_{22} = (d) + (h) = 0.4 + 1.5 = 1.9$

For convenience, let us display this payoff matrix in tabular form:

Player I's payoff matrix:

	Player I maintains confidentiality	Player I breaches confidentiality
Player II maintains confidentiality	1	-0.3
Player II breaches confidentiality	1.2	0.6

Player II's payoff matrix:

	Player I maintains confidentiality	Player I breaches confidentiality
Player II maintains confidentiality	0.3	0.4
Player II breaches confidentiality	0.1	1.9

The game has two possible equilibria. In the case of the equilibrium where both players choose to maintain confidentiality, neither of them individually can benefit from changing their strategy. The same goes for the other equilibrium – both players choosing to breach confidentiality. According to David Lewis, in situations like this, we see a shift from a zero-sum game to a coordination game.[581] This can be achieved by switching from a game of pure strategies to a game of mixed strategies, as follows:[582]

		$1 - p = 0.12$	$p = 0.88$
		Player I maintains confidentiality	Player I breaches confidentiality
$1 - q = 0.74$	Player II maintains confidentiality	1, 0.3	-0.3, 0.4
$q = 0.26$	Player II breaches confidentiality	1.2, 0.1	0.6, 1.9

---

[581] David Lewis, *Convention: A Philosophical Study* (Cambridge: Harvard University Press, 1969), 5-51.

[582] Aumann, Tauman and Zamir, *Torat ha-miskhakim*, 134-139.

Let $0 \leq p \leq 1$ be the probability of Player I choosing to breach confidentiality, and $1 - p$ the probability of Player I choosing to maintain confidentiality. Respectively, let $0 \leq q \leq 1$ be the probability of Player II choosing to breach confidentiality, and $1 - q$ the probability of Player II choosing to maintain confidentiality. p and q can be calculated using the following formulae:

$$p \cdot (-0.3 + 0.6) = (1 - p) \cdot 1 + (1 - p) \cdot 1.2 \rightarrow p \cdot 0.3 = 2.2 - p \cdot 2.2 \rightarrow p = 0.88, \; (1 - p) = 0.12$$

$$q \cdot (0.1 + 1.9) = (1 - q) \cdot 0.3 + (1 - q) \cdot 0.4 \rightarrow q \cdot 2 = 0.7 - q \cdot 0.7 \rightarrow q = 0.26, \; (1 - q) = 0.74$$

Now, we can calculate the expected utility for both players using this formula:

$$\sum_{i=1}^{2} \sum_{j=1}^{2} Aij$$

The expected utility for Player I:

$\sum_{i=1}^{2} \sum_{j=1}^{2} Aij = 0.88 \cdot 0.26 \cdot (-0.3 + 0.6) + 0.12 \cdot 0.74 \cdot (1 + 1.2) = 0.27$

The expected utility for Player II:

$\sum_{i=1}^{2} \sum_{j=1}^{2} Aij = 0.88 \cdot 0.26 \cdot (0.4 + 1.9) + 0.12 \cdot 0.74 \cdot (0.3 + 0.1) = 0.56$

Both players' utility in a mixed strategy Nash equilibrium is smaller compared to what they will get if they stick to their pure strategy. **Player I** gets a utility of *1* if both players maintain the confidentiality of the information, which means a loss of *0.73* in a switch to a mixed strategy. **Player II** gets a utility of *1.9* if both players breach confidentiality, which means a loss of *1.34* in mixed strategy. Lewis explains that the reason players choose to play a coordination game is their expectation that this is the behavior expected of them.[583] In our example, the players will violate confidentiality because they believe that "this is what everyone does" and that individuals are not actually expecting anyone besides themselves to protect the confidentiality of their information.

Appendix A – Summary and Conclusions

In the first two game models discussed here, Player II has a dominant strategy that supports the violation of the privacy of Player I (an individual). This accurately reflects the current state of events in the field of Internet privacy. To improve this situation, either and/or all of the following should happen: one possible solution is more severe fining of confidants and Internet companies who violate personal privacy; this is basically the purpose of the GDPR. However, this is not enough; the paradigm of personal privacy that is founded on consent and control[584] should be replaced by a new one that would pass the responsibility for privacy protection from the individuals, who are expected to give consent and have control over their personal information, to the Internet companies and confidants, who should be held responsible and liable for fair and proper use of the said information.[585] Another solution, which will be discussed in Appendix B, is putting a price tag on personal information[586] and imposing the costs on Internet companies to restrain

---

[583] Lewis, *Convention: A Philosophical Study*, 5-51.

[584] See discussion in Chapter 11.

[585] See discussion in Chapter 11.

[586] See my proposition in Chapter 11 to tax Internet companies based on

their uncontrollable appetite for using personal information as a free resource from which they can profit.

The third model discussed here should be harder to manage financially because Internet companies, which are practically monopolies, have the most to gain from violating our personal privacy. The best way to handle this situation is by stricter regulations and social norms.

---

the amount (in kilobytes) of personal privacy information they use.

Appendix B

Economic Models of Using Personal Information as an Economic Resource

Putting a Price on Customers' Personal Information

Let there be a market where a network app is being sold by a monopoly (e.g., Facebook). Let us assume (without loss of generality) that the app production cost for the monopoly is *0* and that the price customers are willing to pay for it is distributed uniformly between *[0,1]*. Provided the monopoly has no knowledge of how much customers are willing to pay for the app (the value of the app for the customers), it sets (with a probability of *p*) a price of *p*, giving us the following price function:[587]

$$p \cdot (1-p) = p - p^2$$

If we derive this function and set it to *0* in order to get the maximum value of *p*, we will get *1 − 2p = 0*. This means that the monopoly sets a price of $\frac{1}{2}$ for a unit of the app, the monopoly's cumulative profit is $\frac{1}{4}$, and the upper median of customers get cumulative discounts at a value of $\frac{1}{8}$.

Now, let us assume that every customer possesses information that has a direct impact on the price they will be willing to pay for the

---

[587] Maschler, Zamir and Solan, *Game Theory*, 379-380.

app (e.g., a list of people they want to be Facebook friends with) and that the monopoly is willing to pay $r \geq 0$ to any customer who will share that information. Any customer i who shares the information will be offered the app at a customized price of p_i. The rest of the customers will be offered the standard price p.

Based on these assumptions and conditions, we can define a game whose players are the monopoly, customers who share personal information, and customers who do not share personal information. It is a game with incomplete information and a common prior,[588] which means a game where each player has only partial knowledge of the game data. In our case, the customers do not know in advance what customized price they will be offered, and the monopoly does not know in advance the utility u_i of each customer. Still, it does know the distribution of customer preferences.

Description of the Game

The monopoly offers customers a discount in exchange for sharing their information. Let it be $d_i = r \cdot v_i$ (v_i being the amount of information shared by customer i).

- Every customer may decide whether to share personal information and how much of it to share;
- The monopoly is offering standard price p to any customer j who does not share personal information and customized price p_i^m to any customer i who does.

---

[588] Ibid., 319-334.

Therefore, the payoff for customers who do not share personal information is $-p$, and the payoff for the monopoly in this case is $p_m = p$.

Let the price offered by the monopoly to customers who do share personal information be $p_i^m(v_i) = v_i$; now, there are two options of payoff: if $r = 0$, the payoff for the customer will be $p_i = v_i$. If $r > 0$, the payoff will be $p_i = v_i + d_i$. The payoff for the monopoly will be either $p_i^m(v_i) = v_i$, or $p_i^m(v_i) = v_i - d_i$, respectively. This gives us the following game tree:

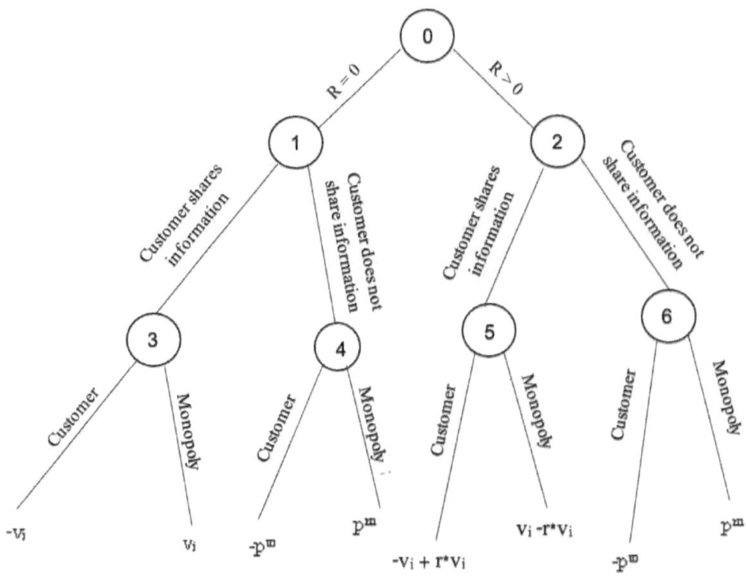

Let us analyze this game according to the Harsanyi model:[589]

The customer possesses two pieces of information: whether $r = 0$ or $r > 0$, and their personal assessment of the utility of the app for them.

The monopoly possesses two pieces of information: whether the customer is sharing their personal information and the distribution of

589 Ibid., 345-346.

customer preferences (the distribution of v). It also knows that the customers who do not share their preferences know the v_i prices, so to protect the monopoly's interest, it is necessary that $p \geq v_i$. Therefore, we can determine that eventually $p^m = 1$.

Thus, we get the following payoff matrix:

For the customer:

		Monopoly	
		$r = 0$	$r > 0$
Customer	Shares information	$-v_i$	$-v_i + r * v_i$
	Does not share information	-1	-1

For the monopoly:

		Monopoly	
		$r = 0$	$r > 0$
Customer	Shares information	v_i	$v_i - r * v_i$
	Does not share information.	1	1

This game has a Bayesian equilibrium, which, in this case, is equivalent to a Nash equilibrium.[590] For the customers, the best strategy is to share their information to secure a price of $v_i \leq 1$, which is lower than the monopoly price $p^m = 1$. Knowing that, the monopoly will always choose $r = 0$ to make sure it gets paid at least $v_i > v_i - r \cdot v_i$. This state of

---

[590] Ibid., 353-356.

events reflects typical customer interaction with monopolistic Internet companies, where individual customers provide their personal information to the Internet companies for free. It also reflects the state of the market, where the monopolies (Internet companies) get all the surplus (more than 25% above what they would have earned without the partial knowledge of customer preferences), while the customers get no surplus at all and have to pay either the full price or the exact price they were willing to pay anyway. In this model, the sharing of personal information is good for the Internet companies and bad for the customers. Furthermore, today, the use of AI technology usually allows Internet companies to obtain personal information without even having to ask for the individual's permission.

Commercializing Customers' Personal Information

An online advertiser has to align their product's price with the customers' preferences. Let *v* be the value of the advertised product unit, distributed uniformly *v = [0,1]*. Let the advertiser's matching function be *m(x)* $= \frac{x^2}{2}$, with *x* standing for the cost of customer communication channels.[591] This means the advertiser's profit from an advertised product is *f(v,x) = vx -* $\frac{x^2}{2}$.

Let there be a database manager who can provide the advertiser with the customer data it needs for a price of p for each customer whose personal data the advertiser wants to buy. Suppose the advertiser can obtain the personal data it needs without buying it from the database

---

[591] Includes costs of website maintenance and online advertising (e.g., through Google Ads).

manager. In that case, it can create a customized advertising package where $x^* = v_i$ for every customer i, to achieve the optimal profit of $f^*(v_i, x^*) = v_i \cdot v_i - \frac{v_i^2}{2} = \frac{v_i^2}{2}$. For the rest of the customers whose personal information is not in their possession, the advertiser has to set a standard price of $x^\wedge$. Now, it has to pick a category of customers whose personal data it will buy from the database manager. Otherwise, it faces a risk of over-advertising or under-advertising. In a state of uniform distribution between $[0,1]$, the equilibrium strategy will be: $v = p = \frac{1}{2}$.[592]

Because v_i is unknown to the advertiser, it can charge customers the price of their data $- p$. If the customer buys the product for p, the advertiser's matching function will be $m(v_i) = v_i^2 - p$. The break-even point is $v_i^2 - p = 0$, from which derives $v_i = \sqrt{p}$. In this case, the equilibrium strategy will be:

$$F^*(v_i) \begin{cases} v_i \geq \sqrt{p} & \text{Buy from the ad} \\ \\ v_i < \sqrt{p} & \text{Do not buy from the ad} \end{cases}$$

The probability of a customer buying the product from the ad is $(p - 1)$, based on the sealed-bid auction with entry fee model.[593] Thus, we will get:

$$F^*(p) = \frac{[(1 - \sqrt{p}) \cdot (4p + \sqrt{p}) + 1}{3}$$

This is a concave function of p in which $F^*(0) = \frac{1}{3}$ and $F^*(1) = 0$. If we derive it, we will get:

$$F'(p) = [(1 - \sqrt{p}) \cdot (4p + \sqrt{p}) + (1 + 3p - 4p^{3/2})]' = 3 - 6 \cdot 1 = \sqrt{p}$$

---

[592] See justification in the previous section.
[593] Maschler, Zamir and Solan, *Game Theory*, 482-484.

The maximum point of the function is $p* = \frac{1}{4}$. Therefore, the price for customers whose $v_i < \frac{1}{2}$ will be: $p* = \frac{1}{2} - \frac{1}{4} = p^l = \frac{1}{4} - \frac{1}{2}$ while the price for customers whose $v_i \geq \frac{1}{2}$ will be: $p^h = \frac{3}{4}$.

In this state of events, the advertiser's expected profit is:

$$F(\tfrac{1}{4}) = \frac{1}{3} \cdot (1 - \frac{1}{2}) \cdot (4 \cdot \frac{1}{4} + \frac{1}{2} + 1) = \frac{1}{3} \cdot \frac{1}{2} \cdot \frac{5}{2} = \frac{5}{12} > \frac{1}{3}$$

This will make the database manager want to maximize p, based on $p \cdot (1 - 2 \cdot \sqrt{p})$. If we derive this function, we will get: $1 = 6 \cdot \sqrt{p}$; therefore, for the database manager $p^d = \frac{1}{36}$.

This model shows that online advertising can guarantee advertisers an expected profit of $> \frac{1}{3}$, and that database managers can secure a revenue of $\frac{1}{16}$ for each customer whose personal data they sell. In other words, this model justifies surveillance capitalism practices of making money by using the personal information of online customers, which has been obtained for zero or near zero price.

To conclude, in this appendix, I described two game theory models that illustrate how our personal information can be (and is being) used to maximize the profits of Internet companies, which exploit it for purposes of targeted advertising and niche (long tail) production.

Appendix C

Technological and Logical Obstacles to

Privacy Protection

In Chapter 11, we reviewed the cracks in all the main conceptions underlying the current paradigm of personal privacy. One of them is the inability to protect personal privacy in the sense of limited access to the self, as defined by Ruth Gavison.[594] The right to limit access to oneself is constantly threatened in the age of ICTs: we may be under surveillance by IoT devices, we are being profiled, we can be identified in public spaces, and the information collected about us makes it possible to get to (or at) us physically, by phone, text or email. All three elements of limited access listed by Gavison – secrecy, anonymity and solitude – are becoming increasingly harder to achieve despite various technological efforts to manage the challenges. In this appendix, I will review the inherent technological flaws of the most common privacy protection methods:

K-Anonymity

K-anonymity is a formal model designed to protect the anonymity of users whose data are stored in a data pool(s).[595] The basic concept of *k*-

---

[594] Gavison, "Privacy and the Limits of Law".
[595] Sweeney, "*k*-Anonymity: A Model for Protecting Privacy".

anonymity is that any given individual whose information is contained in a data release should be indistinguishable from at least *k-1* other individuals whose data also appear in the release. For example, *k = 2* means that a given individual's data cannot be distinguished from the data of at least one other person. The formal definition of *k*-anonymity:[596]

- Let *T* be a table that contains personal data;
- Let *RT* be a table derived from *T*;
- Let $A_1, A_2, ..., A_n$ be a list of attributes;
- Let $a_1, a_2, ..., a_l$ be quasi-identifiers (QI) included in the list of attributes so that for every i, $A_i = a_i$ and $l < n$;

Database *T* can be said to satisfy *k*-anonymity if and only if each sequence of values in Table *RT* appears with at least k occurrences in *RT*.

Naturally, some attributes are unique identifiers, such as ID number, phone number, and lately – biometric data. All these are components of our passport identity.[597] Other attributes are quasi-identifiers, i.e., they are associated with a specific person but are not unique to that person, such as address, date of birth, weight, education, etc. There are several ways to modify table *T*, which explicitly contains all the attributes of a specific person, so that it would satisfy *k*-anonymity, including removing unique identifiers, shortening quasi-identifiers or replacing them with ** characters, etc. Let us consider the following example:[598]

---

[596] Ibid., 6-7.

[597] See definition in Chapter 6.

[598] Sweeney, "*k*-Anonymity: A Model for Protecting Privacy", 9.

	Race	Birth	Gender	ZIP	Problem
t1	Black	1965	m	0214*	short breath
t2	Black	1965	m	0214*	chest pain
t3	Black	1965	f	0213*	hypertension
t4	Black	1965	f	0213*	hypertension
t5	Black	1964	f	0213*	obesity
t6	Black	1964	f	0213*	chest pain
t7	White	1964	m	0213*	chest pain
t8	White	1964	m	0213*	obesity
t9	White	1964	m	0213*	short breath
t10	White	1967	m	0213*	chest pain
t11	White	1967	m	0213*	chest pain

Figure 2 Example of *k*-anonymity, where *k*=2 and QI={*Race, Birth, Gender, ZIP*}

The most important point is that the method of *k*-anonymity only prevents distinction between individual people (identification) based on their quasi-identifiers. The individuals' sensitive attributes remain unscrambled by definition to keep the data meaningful, and that is exactly one of this method's weaknesses. The table in the above example satisfies $k = 2$ regarding the quasi-identifiers but fails to satisfy *k*-anonymity if the Problem column is considered as well: for example, *t1* has no equivalent.

In practice, maintaining privacy by making sure data releases satisfy *k*-anonymity is failing for the following reasons:

- The underlying assumption is that databases are separate and cannot be combined. However, in practice, there is almost always a way to fuse databases and thus nullify the anonymity. Let us take, for example, the following private table *PT* and linked table *LT*:[599]

---

[599] Ibid., 11.

If we combine the two tables, everybody's full dates of birth will be revealed, breaching the $k = 2$ anonymity of *LT*.

<table>
<tr><td>

Race	BirthDate	Gender	ZIP	Problem
black	9/20/1965	male	02141	short of breath
black	2/14/1965	male	02141	chest pain
black	10/23/1965	female	02138	painful eye
black	8/24/1965	female	02138	wheezing
black	11/7/1964	female	02138	obesity
black	12/1/1964	female	02138	chest pain
white	10/23/1964	male	02138	short of breath
white	3/15/1965	female	02139	hypertension
white	8/13/1964	male	02139	obesity
white	5/5/1964	male	02139	fever
white	2/13/1967	male	02138	vomiting
white	3/21/1967	male	02138	back pain

PT

</td><td>

Race	BirthDate	Gender	ZIP	Problem
black	1965	male	02141	short of breath
black	1965	male	02141	chest pain
black	1965	female	02138	painful eye
black	1965	female	02138	wheezing
black	1964	female	02138	obesity
black	1964	female	02138	chest pain
white	1964	male	02138	short of breath
white	1965	female	02139	hypertension
white	1964	male	02139	obesity
white	1964	male	02139	fever
white	1967	male	02138	vomiting
white	1967	male	02138	back pain

LT

</td></tr>
</table>

- Another way to breach the anonymity of a data release is by having prior knowledge of the problem or the attributes of the person in question. For example, if we look at the following medical release:

DSM diagnosis	Gender	ZIP	Date of birth
HIV	*	Center	199*
HIV	*	Center	199*
Bulimia nervosa	*	Center	199*
Bulimia nervosa	*	Center	199*

Assuming that the ratio of males to females is 50:50, and knowing that the prevalence of HIV in young men (under

353

35) living in the Central region is 20%, while the prevalence of bulimia nervosa is 2% if we find out that Dan – a man in his 20s living in the Central region – has a medical condition, we can infer with a high level of certainty that he has HIV.

Conclusion: Because the vast majority of data releases that contain personal information are included in at least one Internet database, and the people who analyze them usually have some background knowledge that helps them overcome the quasi-identifier scrambling, there is almost always a way to breach k-anonymity and violate one's privacy by combining different databases and using prior knowledge.

L-Diversity

In the previous section, I reviewed the inherent flaws of k-anonymity and showed how it fails to guarantee anonymity (privacy) of data in large data releases due to insufficient protection of the sensitive data. This understanding gave birth to the principle of l-diversity – an extension of k-anonymity that measures the diversity of sensitive values for each column in which they occur. According to the formal definition, a data release has l-diversity if, for every set of rows with identical quasi-identifiers ("equivalence class" or "block"), there are at least l well-represented values for each sensitive attribute.[600] For example, the table

---

[600] Ashwin Machanavajjhala, Johannes Gehrke and Daniel Kifer, "ℓ-Diversity: Privacy Beyond k-Anonymity", 22nd International Conference on Data Engineering (ICDE'06), 3-7 April 2006, 2.

below has *k*-anonymity of $k = 4$ but no *l*-diversity in the third block (lines 9-12), where the medical condition is the same for all four entries.

	Non-Sensitive			Sensitive
	Zip Code	Age	Nationality	Condition
1	130**	< 30	*	Heart Disease
2	130**	< 30	*	Heart Disease
3	130**	< 30	*	Viral Infection
4	130**	< 30	*	Viral Infection
5	1485*	≥ 40	*	Cancer
6	1485*	≥ 40	*	Heart Disease
7	1485*	≥ 40	*	Viral Infection
8	1485*	≥ 40	*	Viral Infection
9	130**	3*	*	Cancer
10	130**	3*	*	Cancer
11	130**	3*	*	Cancer
12	130**	3*	*	Cancer

However, *l*-diversity cannot help protect data anonymity if there is insufficient diversity in the sensitive attribute to begin with, i.e., if there is insufficient distribution of the sensitive attribute in a given population or if *l*-diversifying the dataset compromises the integrity of the data. Let us consider, for example, the following table:

Date of Brith	Zip	HIV (sensitive data)
9*	Center	Negative
9*	Center	Negative
9*	Center	Negative
9*	Center	Negative
8*	North	Negative
8*	North	Negative
8*	North	Negative
8*	North	Positive

In this example, *l*-diversity cannot provide adequate protection of personal privacy because there is only one person from the North region born in the 1980s who is HIV-positive, which means that person can be easily identified. To meet the requirement of $l = 2$, we would have to change the "HIV" value in one of the lines 5-7 from "negative" to "positive," but that would misrepresent all the people in that class as 50% HIV-positive (even though the actual prevalence in the general population is 1%), presenting a serious risk to their privacy.

T-Closeness

The principle of *t*-closeness was designed to address the limitations of *l*-diversity.[601] The basic requirement is that the distribution of a sensitive attribute in any equivalence class be close to the distribution of the attribute in the overall table before it was *k*-anonymized. According to the formal definition, an equivalence class is said to have *t*-closeness if the distance between the distribution of a sensitive attribute in this class and the distribution of the attribute in the whole table is no more than a threshold *t*. A table is said to have *t*-closeness if all equivalence classes have *t*-closeness.[602]

All three abovementioned principles are based on the redaction of unique identifiers and scrambling of quasi-identifiers and can help protect the privacy of personal information provided that an individual cannot be identified based on their quasi-identifiers alone.

---

[601] Ninghui Li, Tiancheng Li and Suresh Venkatasubramanian, "*t*-Closeness: Privacy Beyond *k*-Anonymity and *l*-Diversity", 2007 IEEE 23rd International Conference on Data Engineering, 5-20 April 2007.
[602] Ibid., 4.

One of the known limitations is that making semantic changes to a database might lead to information loss and, therefore, compromise the usefulness of the data release: for example, if the anonymization process changes the correlation between the distribution of symptoms and the distribution of medical conditions, the data release cannot be used for medical inference.

Another limitation is the false assumption of lack of background knowledge. Most people have extensive background knowledge, which can be used to narrow down the possibilities and, when combined with AI algorithms, overcome these methods of personal privacy protection.

Differential Privacy

Differential privacy (DP) is a mathematical framework designed to address the inherent limitations of the information privacy protection methods discussed previously in this appendix.[603] It can provide a strong guarantee of privacy by allowing data to be analyzed without revealing sensitive information about any individual in the dataset.[604]

According to Kobbi Nissim, the intuitive idea that removing identifiers guarantees the protection of information privacy is often wrong. Science and practice have proven the routinely applied statistical

---

[603] Nissim, "Differential privacy: Why, How, and Where to?".
[604]https://www.statice.ai/post/what-is-differential-privacy-definition-mechanisms-
examples#:~:text=Differential%20privacy%20is%20a%20mathematical,any%20individual%20in%20the%20dataset

tools and methods of informational privacy protection (illustrated by the figure below) to provide less than sufficient privacy protection.[605]

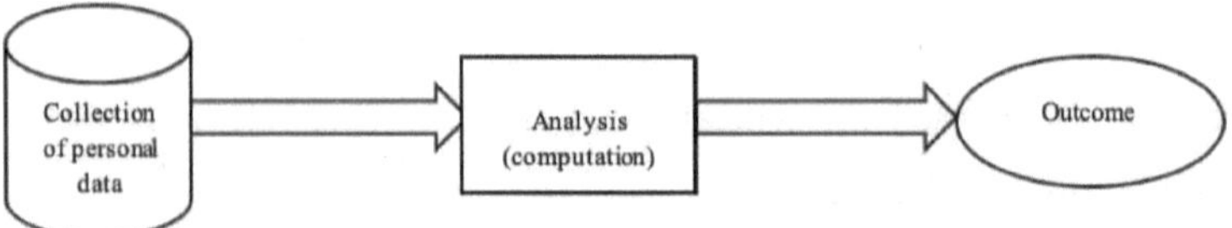

Nissim gives the following example:[606] NIH collects DNA from people with a certain medical condition and publishes minor allele frequencies for the control group, which includes 100,000 people.

5%	3%	12%							3%

Given that the frequencies in the overall population are:

3%	3%	11%							3%

If an individual's genome is:

1	0	0							1

We can infer with a high level of certainty that the person is not in the control group.

According to Nissim, it is a mathematical fact, regardless of privacy policy, that any useful analysis of personal data must leak some information about individuals and that the leakages accumulate with

---

[605] Nissim, "Differential privacy: Why, How, and Where to?", 8-15.
[606] Ibid., 16.

multiple analyses/releases.[607] That is true, he argues, for any statistical privacy protection method.

The purpose of differential privacy is to address this issue. Differential privacy means that sensitive personal information is slightly modified by the addition of carefully crafted random "noise" so that the inclusion of the said personal information in the analysis does not present a privacy threat. The analysis is said to satisfy differential privacy if the inclusion or non-inclusion of one's personal information does not affect one's privacy, e.g., one's chances of getting arrested, denied insurance, shamed socially, etc.[608]

Thus, differential privacy can guarantee adequate protection of information privacy if the computation (analysis) process meets the requirements. However, it does not provide adequate protection in the following cases:

- Differential privacy usually cannot protect against the invasion of deep personal privacy.[609] Violations of deep personal privacy often make use of indirect information, which means it does not matter whether one's data is included in the dataset or not.

- Differential privacy is not applicable when one has to provide specific information that needs to be acted upon – for example, when one makes an online transaction or surfs the web – so it cannot prevent profiling and use of personal information for profit by the Internet companies. Moreover, it cannot protect

---

[607] Ibid., 21.
[608] Ibid., 41.
[609] See Chapter 8.

individuals who are nodes in the Internet network against invasion of their deep and general personal privacy.

Despite these limitations, differential privacy has many applications;[610] among other things, it can be used in data mining to balance the protection of information privacy against the accuracy of decision tree induction in data mining.[611]

In this appendix, I have given an overview of today's most common methods of ensuring the privacy of information in big data pools, as well as their strengths and limitations. Even though each of those methods provides some protection for information privacy, they fail to provide an adequate solution to the privacy challenges posed by ICTs. Therefore, I believe, as explained in Chapter 11, that we must look for other solutions in the form of a new paradigm of personal privacy that would be more compatible with the spirit of the age of ICTs.

---

[610] Nissim, "Differential privacy: Why, How, and Where to?", 63-78.

[611] Arik Friedman and Assaf Schuster, "Data Mining with Differential Privacy", delivered at: Privacy in Challenging Times: The 8th Technion Summer School on Cyber and Computer Security (September 2020).

Index of Names

Index of Key Terms

European Union, EU   8, 128, 130, 135, 206, 228, 281, 287, 289, 305, 306, 307, 309, 310
exemplar(s)   123, 133, 218, 277, 286, 317
external entities   47, 50, 105, 145, 146, 160, 175, 182, 263
Facebook   07, 20, 24, 25, 26, 30, 49, 50, 105, 111, 112, 134, 135, 145, 152, 172, 180, 184, 186, 190, 192, 190, 84, 206, 220, 223, 228, 229, 242, 260, 261, 276, 278, 282, 287, 289, 306, 307, 311, 343, 344
Fordism   84, 86, 87, 88, 168, 183, 309
game theory   192, 281, 297, 298, 302, 317, 318, 319, 343, 348, 349
Nash equilibrium   281, 318, 328, 334, 340, 346
payoff matrix(-ces)   322, 323, 325, 326, 327, 328, 331, 332, 333, 336, 338, 339, 346
pure strategy(-ies)   319, 328, 334, 339, 340
general [personal] privacy   9, 12, 13, 120, 137, 138, 139, 140, 142, 144, 146, 237, 263, 283, 288, 290, 293, 294, 319, 320, 321, 328, 334
general [personal] privacy information   140, 231, 294, 320
ghost work   180
Gilder's law   267
global village   28
hacker(s)   7, 29, 289
hump(s)   140, 220, 263, 319, 320, 321
big hump   263, 320, 321
little hump   263, 319, 321
identity   5, 07, 13, 24, 30, 32, 47, 50, 55, 57, 67, 72, 90, 95, 102, 103, 106, 107, 108, 109, 110, 111, 112, 113, 114, 115, 126, 127, 130, 138, 144, 147, 148, 149, 150, 151, 152, 153, 155, 157, 160, 161, 162, 166, 173, 174, 175, 176, 178, 182, 190, 207, 213, 216, 226, 227, 244, 270, 271, 273, 294, 297, 307, 308, 312, 317, 351
attachment identity   109, 111, 149
attribution identity   108, 111, 148
numerical identity   108, 111, 147
offline identity(-ies)   111, 112, 239
online identity(-ies)   109, 110, 111, 112, 113, 127, 153, 226, 260, 270, 271
passport identity   47, 108, 111, 114, 115, 147, 148, 161, 174, 175, 182, 190, 351